Impression Procedures in Prosthodontics

Contributing Authors:

Dr. Lalit Kumar

Senior Assistant Professor,

Dr Harvansh Singh Judge Institute of Dental Sciences & Hospital, Panjab University, Chandigarh.

Dr. Prashant Pujari

Reader,

Pacific Dental College & Hospital,

Udaipur.

Dr. Rakshith Hegde

Associate Professor

AB Shetty Memorial institute of Dental Sciences and Hospital

Mangalore

Dr. Charu

B.D.S,

Dental Officer, NRHM

Chandigarh

CONTENTS

1. **Introduction**.. 1
2. **Classification of impression materials**... 6
3. **Impression procedures**
    a. *Impression procedure for RPD*... 9
    - Impression trays... 11
    - Initial impression procedure.. 13
    - Final impression procedures.. 21
        - *Altered cast procedure*.. 24
        - *Functional reline procedure*.. 27
        - *Sectional impression*.. 28
        - *Fluid Wax technique*.. 29
        - *Functional Impression*.. 34
        - *Static reline technique*... 36
        - *Functional Relining Technique*................................... 37
        - *Hindel's technique*... 39

    b. **Impression procedure for FPD**
    - Initial impression procedure.. 41
    - Tray fabrication.. 43
    - Tissue displacement.. 44
    - Final impression procedure... 50
        - Hydrocolloid.. 50
        - Poly-sulphide... 57
        - Silicone.. 59
        - Polyvinyl siloxane.. 61
        - Polyether... 62
        - Dual arch impression... 63
        - Double arch impression... 66
        - Combination impression.. 69
        - Impression; with bite registration................................. 71
        - Impression for pin retained restorations....................... 73

4. **Bibliography**.. 75

INTRODUCTION

According to Glossary of Prosthodontics Terms[28], an impression is defined as *"A negative likeness or copy in reverse of the surface of an object; an imprint of the teeth and adjacent structures for use in dentistry"*.

A preliminary impression is a negative likeness made for the purpose of diagnosis treatment planning or the fabrication of a tray. According to *Shillingburg*[72], an impression is an imprint or negative likeness. It is made by placing some soft, semi-fluid material in the mouth and allowing the material to set.

According to *DeVan*[17], an impression is a registration of mouth tissues with an impression material. It is a record, a facsimile of mouth tissue taken at unstrained rest position or in various positions of displacement.

The partially edentulous condition can be restored with removable partial dentures or fixed partial denture. The ultimate goal of prosthodontic service is the placement and maintenance of prosthesis, which are in biologic and functional harmony with the supporting tissues and remaining teeth.

Historically, various materials have been used to make impression for fixed prosthodontics, *Phillip Pfaff*[88] in Berlin was the first with a wax impression from the jaw and plaster casting from it for prosthetic purpose. Early materials included rigid and semi rigid compositions such as plaster zinc-oxide eugenol, compound and waxes, these materials still have limited use in dentistry. However, current fixed prosthodontic impression are the domain of the elastic materials[66], including reversible hydrocolloid and four types of synthetic elastomers.

The method used to make impression of supporting and retaining anatomic structure of the mouth is of basic importance for obtaining optimum distribution of the masticatory load in the construction of removable partial denture.[31] A partial denture constructed on a cast made from an impression which does not fulfill this requirement will be a failure regardless of how well designed and executed it might be.[35] A widely accepted axiom of removable partial prosthodontics holds that it is fully as important that the denture be designed and constructed in such a way as best to preserve the oral structures, as it is to restore function. The principal functional basing is employed to create conditions, which favour maximum longevity of the remaining structures.

Partial edentulous mouths with distal extension edges present the challenge of correctly registering two tissues as dissimilar as teeth and edentulous ridges[38, 48]. Obtaining the impression of the removable partial denture can be a thoroughly unpleasant experience, from the patient's point of view, if it is not accomplished with skill and finesse. The dentist who prepares himself with an intimate knowledge of impression material with which he is working and who follows a technique which is consistence with its physical properties, can eliminate most of the unpleasant aspects for all but the most squeamish the patients.

To achieve a successful impression, the following concepts should be adhered to:

1. The impression material must become elastic after placement in the mouth because it must be withdrawn from under-cut regions that usually exist on the external tooth surfaces.

2. It must have adequate strength to resists breaking or tearing on removal from the mouth

3. It must have adequate dimensional accuracy, stability and reproduction of details so that it is an exact imprint of prepared and adjacent unprepared teeth.

4. It must have handling and setting characteristics that meet clinical requirements.

5. It must be free of toxic or irritating components

6. It should be exact duplication of the prepared tooth, including all of the preparation and enough uncut tooth surface beyond preparation to allow the dentist and technician to be certain of location and configuration of finish line.

7. Other teeth and tissue adjacent to prepare tooth must be accurately reproduced to permit proper articulation of cast

8. It must be free of bubbles especially in areas of finish line and occlusal surfaces of the other teeth in the arch.

A variety of techniques has evolved overtime. Selection of the specific techniques depends on experience and an evaluation of an individual patient. Time, expense and accuracy must all the considered in making the selection.

A thorough understanding of the characteristics of each of the impression materials leads to the obvious conclusion that no single material can record both the anatomic form of teeth and tissues in the dental arch and at the same time the functional form of the residual ridge.

Impression should be perfectly accurate and removed with care. Although, it is practically impossible to determine exactly the magnitude of the force that oppose the withdrawal of impression, it is possible to form an idea of the external and internal forces involved and the possible influence of the forces upon impression materials.[87] The impression must be handles properly until it is poured up in a gypsum product. The vital portion of the impression is poured first.[55] As a general rule, an appropriate disinfections protocol should be followed after impression removal from the mouth

A detailed protocol is required to insure that the impression is handled properly. The impression must be rendered harmless before being passed on to other people who will work with it or with the gypsum cast made from it, outside the dental operatory. Different types of chemical disinfectants are available for this purpose.[72] A prolonged immersion in 2% glutaraldehyde or hypochlorite solution with 10,000 ppm available chlorine for 1 hour was recommended for disinfecting impressions in a 1973 World Health Organization report. A later recommendation for the disinfecting of impressions and casts published by the ADA Council on Dental Material, Instruments and Equipment; Council on Dental Practice, Council on Dental Therapeutics in 1988 and amended in august 1991 call for immersion of polysulfide, condensation-reaction silicone, polyvinyl siloxane, polyether and agar hydrocolloid in ADA accepted disinfecting solutions that require immersion for no longer than 30 minutes. An alternative technique, spraying, can be used on those materials most vulnerable to distortion.

CLASSIFICATION OF IMPRESSION MATERIALS

Impression materials can be classified as:

1. **Rigid materials**

    a. Plaster of Paris

    b. Metallic oxide pastes

2. **Thermoplastic materials**

    a. Modeling plastic

    b. Impression waxes and natural resins

3. **Elastic materials**

    a. Reversible hydrocolloid (agar-agar)

    b. Irreversible hydrocolloid (alginate)

    c. Mercaptan rubber-base impression materials

    d. Silicone impression materials

    e. Polyethers

Impression techniques [11] can be classified as:

1. According to the position of the jaws in relation to each other when the impressions are being made.

    a. Closed mouth

    b. Open mouth

2. According to the degree of pressure applied when the impression are made

    a. Positive pressure

    b. Negative pressure

    c. Selective pressure

    d. Non pressure

IMPRESSION PROCEDURES FOR REMOVABLE PARTIAL DENTURE

Removable partial dentures can be broadly divided into two groups.

- Tooth supported and
- Tooth – tissue supported.

Support for tooth mucosa supported removable partial dentures is derived from structures with markedly different viscoelastic responses to loading[18] It is derived from a combination ;of hard tooth structure and soft mucosal tissues with their underlying bony support[2].

It has been stated, "Removable partial dentures which receive their support from teeth and edentulous areas should have that support as equally divided as possible. Equalization of stresses can be obtained by using a secondary impression of the base and supporting tissues."

McLean stated, "The basic problem of partial denture stabilization is to equalize the resilient and non resilient support". [57]

The form and relationship of the remaining teeth must be accurately recorded, with the anatomic form of teeth and tissue being registered with exactness. The literature is lacking in reports of investigation of the amount of support that can be obtained by various methods. There is no consensus regarding the best method for resolving this problem. Denture base movement resulting from an occlusal load is influenced by impression material and impression technique used to obtain support for partial denture.

The philosophies exist on the residual ridge form that is best obtained in impression for favorable denture base support. Some clinician advocate making impression with no pressure applied to the residual ridge. Whereas others advocate loading the residual ridge during eh impression procedure. Many investigators feel that a functional impression technique is one of the major steps the dentist can take to achieve the most favourable prognosis for the distal extension removable partial denture.

The supporting structure of the edentulous ridge and the remaining teeth are utilized to best advantage by altering the edentulous part of the cast upon which a distal extension removable partial denture is fabricated. A number of reports have proposed procedures for registering jaw relation and making the altered cast impression during the same visit.[1]

IMPRESSION TRAYS

For the purpose of conveying the impression material into the mouth and over the teeth, and holding it in this position until it hardens, specially designed impression trays are customarily used. Preliminary impressions are used as the first step in the construction of a final impression tray. [58]

Impression trays can be classified broadly in two:

- *stock trays* and
- *custom trays*.

Stock trays are made by the dental manufacturers, most commonly of metal, in a variety of sizes to fit large, medium and small mouths. Stock trays may be dentulous for use with the dental arch which has some remaining natural teeth, or edentulous for the mouth with no remaining teeth. Another category is the depressed anterior tray, designed especially to be used for the mouth, which has only anterior teeth remaining.

Stock trays for partial dentures may be perforated to retain the impression material in place better, or they may be constructed with a rim lock for this purpose. The rim lock retains the impression material in position in the tray by means of an overlapping edge, which wedges it in place. Another type of stock trays, designed for use with reversible type of hydrocolloid, is the water-cooled tray. It contains tubes through which water can be circulated for the purpose of cooling the agar.

Custom trays are sometimes needed for mouths that are abnormally large or small or of unusual configuration. Another indication for the custom tray is the ease where all of the peripheral borders must be precisely delineated in the impression. The custom tray is especially recommended for the impression in which it is desirable to establish precise borders. The custom tray may be made of acrylic resin or shellac base plate material. An acrylic resin custom tray can provide a uniform and specific thickness of impression material and result in improve accuracy [25]. The thickness is controlled in order to minimize the dimensional change during polymerization of elastomers.

The modified stock tray (individualized tray)

The stock tray can be modified with modeling composition or with wax to create, a very accurately fitted tray, which, for want of a better name and to distinguish it from the conventional custom tray, may be termed "individualized tray". The prime indication for the individualized tray is the mouth with edentulous spaces, which are not tooth, bounded the Kennedy Class I and II in particular.

Technique for Tray Fabrication

Softened modeling composition is placed in the stock impression tray in such a way that it may capture the edentulous areas of the mouth and include one or two teeth adjacent to the space. The tray is positioned in the mouth and the compound is allowed o cool, but it is not permitted to harden completely, so that it is prevented from becoming locked around the adjacent teeth. When it has hardened sufficiently to maintain its contour, it is removed from the mouth and thoroughly chilled. The compound is trimmed so that it does not contact the adjacent teeth, and the surface of the compound in the edentulous areas is scraped to a depth of 2 to 4 mm., to provide space for a uniform layer of impression, the compound should cover the edentulous ridge and the palate, and should be accurately fitted to the post dam area.

Wax may be employed to create a customized tray. The individualized tray technique is especially useful for the mouth that is either exceptionally large or small, or the one of anomalous contour which cannot be accurately fitted with a conventional stock tray.

INITIAL IMPRESSION PROCEDURE

Patient Management

The impression procedure, besides engendering the trepidation of a new experience, may activate a subconscious fear of having the airway blocked. This can give rise to a feeling of mild panic in a patient who is already apprehensive. It is time well spent, therefore, to devote a few moments to explaining to the patient, for whom an impression is to be made, that it is basically a simple procedure, and to reassure him that there is no need to feel anxiety. Merely reassuring such an individual that there is nothing in the procedure to cause him more than mild discomfort will usually suffice to relax him and restore his confidence and sense of security.

Positioning the Patient

The patient should be seated comfortably upright, with the head firmly supported in the headrest. The plane of occlusion should be approximately parallel to the floor. He should be instructed to sit quietly and to relax. The average patient who has no nasal obstruction may be instructed to breath through the nose while the impression tray is in place. He should be instructed not to swallow while the impression is in place. He may be advised not to be concerned about the excess saliva that may accumulate in the floor of mouth. Not to be overlooked is the fact that the patient should be suitably draped to protect his clothing from accidental spillage of impression material or saliva

Mouth preparation and prophylaxis

It is not recommended that a prophylaxis be accomplished immediately before obtaining the impression, however, as this will sometimes causes the alginate to stick to the teeth. The prophylaxis should ideally be accomplished 24 hours, or more, before the impression appointment. The soft plaques around the teeth should be removed by cleaning the teeth lightly with pumice immediately prior to making the impression. A nondrying silicone material applied to the tooth surfaces helps to assure the release of the impression material.

Saliva control

Excessive volume can usually be controlled by having the patient rinse with ice water just before the impression tray is inserted, in order to close the orifices of the salivary glands partially. Another approach is to place gauze packs or cotton rolls opposite Stensen's ducts, as well as in the floor of the mouth under the tongue, just before mixing the impression material. The packs should than be removed immediately before inserting the impression tray

The mucinous type of saliva can usually be controlled by means of mouthwash consisting of ½ teaspoon of bicarbonate of soda in a half of a glass of water. The pre-impression rinse has a thinning effect on the saliva, so that it is much less likely to obliterate tissue detail by intervening at the impression tissues interface. If a mouthwash is not at hand, the problem may be overcome by employing the "tanden" impression technique, in which one impression is taken to "soak up" the bubbles and mucinous saliva, followed immediately by a second impression which will record the tissues in a relatively saliva free state.

Tray selection

The tray should be moistened with water before trying it in the mouth, so as to reduce to a minimum frictional contact with the lips and oral mucosa. The sides of a properly fitted tray should avoid direct contact with the buccal and labial surfaces of the teeth by approximately ¼ inch. The maxillary tray should extend posteriorly to include the humular notches. The mandibular tray should be extended to include the retromolar pads. When fitting the trays to the mouth, a comfortable holding position should be rehearsed so as to obviate the need for a shift in finger position during the time that the impression is being recorded.

Many dentists used stock impression trays for partially edentulous patients. Either the rim lock or the perforated type of tray may be used. However, selection of the proper size and shape is essential in order to take advantages of the dimensional accuracy of the elastic impression materials, and to insure the inclusion of all essential areas in the impression.

If the tray is too large, tissues may be stretched or distended, and the quantity of impression material may be too great in some places to permit the making of an accurate impression. It the tray is too small, it may impinge upon and distort the soft tissue, or it may cause the impression material to have thin, fragile areas, which distort easily when making the cast.

The correct size and form of tray can be inserted and removed easily with a clearance of approximately ¼ inch in all regions when an alginate impression material is used. Often stock trays may require some alteration to avoid any tissue impingement of the tray for the individual patient.

The fit of stock trays can be improved by adding modeling plastic in the edentulous regions, in the region of the posterior palatal seal, and on the borders.

Some dentists prefer to use custom tray for alginate impression and other use stock trays. A stock rim lock non-perforated tray is recommended for impressions of partially edentulous patients. Alginate impression materials do not adhere to metal trays so perforated trays generally are used to retain these impression materials.

Jordan preferred un-perforated metal trays coated with beeswax in which cotton fibers were attached to retain the impression materials. *Atkinson, Gill and Shephered* reported that the tray retention of alginate impression materials to un-perforated trays coated with sticky molten wax was two to three times greater than the retention to perforated trays. *Skinner and Phillips*[76] suggested that the proper retention of the impression will be affected by the size and number of perforations and their location and distribution in the tray.

The optimum amount of bulk for alginate impression material must be established to obtain the maximum accuracy. a stock tray can be customized simply and adequately by adding modeling plastic in the edentulous areas of the ridges and the palate.[15]

Tray preparation

The fit of the stock tray should be refined by rimming the peripheral borders with some readily moldable wax, such as utility wax. This will include the post dam area of the maxillary tray. The tray should be modified with wax any area of the mouth that does not encompass reasonably well.

If the vault area is unusually deep, the palatal region of the maxillary tray should be built up with utility wax or modeling composition. When wax is added to this area, holes should be punched in the wax at approximately ¼ inch interval. A stock tray can be customized simply and adequately by adding modeling plastic in the edentulous areas of the ridges and in the palate. Sufficient modeling plastic should be used to permit "reading the impression" of the soft tissues.

The tray and modeling plastic should be chilled in cold water and the modeling plastic trimmed with a sharp knife until it is ¼ inch away from the teeth and between 1/8 and ¼ inch away from the soft tissues. The surface of the modeling plastic should be prepared to hold the alginate material by flaming the surface of the modeling plastic and embedding bulk cotton fibers in it.

Proportioning

The use of the pre-weighed volume of powder mixed with the previously determined amount of distilled water will give a mix of the same consistency every time. Distilled water is recommended for use with all alginate hydrocolloid mixes. [68] The setting time can be controlled to a large extent by varying the temperature [32]. The setting time should not be controlled by varying the consistency of the mix.

Spatulation

Vacuum mechanical spatulation of the alginate mix will produce smooth, air free mixes. It requires approximately 15 seconds of mixing with 20 to 28 pounds of vacuum[67]. The use vacuum spatulated mix will tend to avoid entrapment of air bubbles in critical parts of the impression. When hand spatulation is used the alginate should be mixed vigorously for one minute, timed by a clock.

Stenenson[81] suggested a method for better grasp of bowl. A wet rubber bowl is difficult to hold when rubber gloves are worn. Traction strips are added to rubber bowl to permit firmer grasping more vigorous spatulation. For this, 4 or 5 thin, evenly spaced strips or cyanoacrylate glued on the outer surface of the rubber mixing bowl. Before the glue dries, sprinkle dry pumice of any coarseness over the strips. Allow one minute to dry, and then shake off excess pumice. The procedure may need to be repeated periodically.

Fill in the tray

The tray should be dry when the impression material is placed in it. the alginate is best spooned into the tray with a spatula, using a spreading motion against the bottom and sides of the tray, in order to squeeze out air and force the material into the perforations or under the rim lock. In all but unusual cases, a level, full tray is slightly more material than is needed.

Impression sequence

It is usually preferable to obtain the mandibular impression first, all else being equal. One the other hand, when the prosthesis is to be made for the mandibular arch, it may be best to register the impression for the opposing cast (the maxillary first, so that the mandibular cast can be poured immediately after removal of the impression from the mouth.

Inserting the tray

Remove the gauze packs from the mouth and with the finger tip carefully place some of the mix in the palate. Do not rub the alginate over the teeth. Insert the filled tray immediately, gently, jiggling it into place. Do not seat the tray far enough for it to touch the teeth. If the alginate mix is thick enough, it can be held in place easily without pressure and without stops.

It is imperative to release the pressure as soon as the tray is seated. Maintain only enough pressure to hold the tray in place. It is essential to avoid moving the tray until after complete gelation. The impression should be held in place, without movement of any kind, until the material has completely set.

A test sample of alginate mix on the bracket tables is a reliable means for determining when to remove the impression from mouth. Between one and two minutes after the sample mix on the bracket table has lost its tackiness, the impression should be removed with a firm, quick snap.

Mandibular impression

Standing to the right and in front of the patient, he is instructed to open wide, and the mouth tray is inserted sideways into the mouth. With a rotary motion, it is lined up with the area to be registered, with the tray handle approximately parallel to the occlusal plane and in line with the midline. The patient is instructed to close slightly, so as to increase the space in the vestibules, and the tray is guided gently but firmly into place. When the tray is seated, the patient is instructed to lift the tongue and touch the roof of the mouth with it.

Maxillary impression

From a position to the right of, and slightly behind, the patient, he should be instructed o open wide, and the tray is inserted sideways into the mouth. With a rotary motion, it is lined up with the area to be registered, with the tray handle approximately parallel to the occlusal plane and in line with the midline. The patient is instructed to close slightly, so as to increase the space in the vestibules, and the tray is guided gently but firmly into place. .

The patient should be instructed to close slightly from the wide-open position, so as to increase the space in the vestibules, as well as to withdraw the coronoid process from the buccal space. The posterior border of the tray should first be guided into place until the impression material begins to exist from the posterior border. At this juncture the upward pressure should be directed so as to guide the anterior part of the tray up into its seated position. When the tray has been seated, the patient is instructed to lift the tongue and touch the roof of the mouth with it the upward pressure is released and the tray is held with firm, steady pressure until the material has gelled

Removing the impression from the mouth

To remove the tray, a finger of the free hand should be wedged gently between a peripheral border of the impression and the adjacent tissues of the vestibule in the bicuspid molar area, at the same instant the direct pressure is exerted downward and slightly forward (for the maxillary impression). The direction of removal of the mandibular impression should be upward and in slightly labial direction[34]. A sudden blast of air directed at the border of the impression frequently aids the release of the impression. [92]

Inspection of the impression

The impression should be blown off with a gentle stream of air and inspected under a good light. All excess impression material should be trimmed away with the sharp knife

Care of impression

The impression should be rinsed; it should be wrapped in a wet towel preparatory t pouring of the cast. It should not be laid down on a bench top without support, for fear of distorting it. if the impression cannot be poured immediately, it should be secured in such manner that the tray handle will support the impression. If excess saliva remains in the impression, it must be remove by gently flushing the impression with water. Excess water must be removed from the impression before the cast is poured. The diagnostic cast is formed by pouring the initial impression with gypsum product. The custom tray is fabricated over it for making final impression.

Classification:

Techniques of obtaining the impression may be classified according to the method used to record the tissues, into

1. The open mouth method
2. The closed mouth method

The open mouth technique consists in introducing the tray containing the impression material into the mouth and holding it in place until the material has gelled or set.

The closed mouth method, on the other hand, entails placing the impression tray in the mouth and having the patient hold it in place by occluding on it. Usually this technique is employed with a denture that is to be refitted (a reline), or by means of a customized impression tray. If a custom tray is employed, an occlusion rim of modeling composition is attached to the tray upon which the patient can comfortably close as the impression is being registered.

Impression making procedure for distal extension removable partial dentures falls into *two broad general groups*.

The first involves the use of a single mix of impression material to make an impression and a one piece cast to fabricate the prosthesis. The impression is usually made with alginate (irreversible hydrocolloid) impression material in a stock tray, and the basal surface of the denture is formed on the cast obtained from this record of the teeth and soft issues.

The second group involves the use of a two section impression and either a one or a two section cast to fabricate the prosthesis

There are two subdivisions within this group. The proponents of one technique advocates making an impression of the edentulous ridges with the tissue in functional position, and the proponents of the other advocates making the impression with the mucosal tissue at rest.

For example, Hindel's uses a technique for making a impression for producing a two section cast from what he describes as an anatomic, or tissue at rest, impression. Others believe that impression produced by both techniques register a functional relationship between the edentulous and the dentulous region in the dental arch and use an altered cast procedure to obtain a registration with the mucosal tissues at rest [89]

It is sometimes desirable to exploit the advantages of more than one technique or material in an impression by using two different materials in two separate steps[49]. This is referred to as a two-piece impression.

FINAL IMPRESSION PROCEDURE

Preservation of the remaining oral structures in such a state of health as to ensure their longevity is best achieved by exploiting tooth support to the maximum extent that is available. The prosthesis of ideal design is one that is entirely tooth-borne.

Unfortunately, the number and distribution of the teeth remaining in the mouth of the typical candidate for a partial denture makes this not often possible of achievement. When complete tooth support is not available, stress should be distributed equally between the remaining natural teeth and all of the soft tissue support that is capable of bearing a portion of the masticatory load. A problem is encountered in employing this composite type of support, however, because the mucosa, as a result of its resilience and displaceability provides an unstable foundation for the prosthesis. The amount of tissue displacement varies in different patients when using the same impression techniques according to histological characteristics of the tissues.

This is explained by the fact that the oral mucosa in stress bearing areas of the mouth is capable of assuming two distinctly different contours. One is the form which it assumes when it is at rest – the resting or anatomic form; the other is its functional or supporting from, which it assumes when it is subjected to the pressure of an occlusal load. The distortion in the master cast in the vertical direction is related to the accuracy of the impression and the material used in this procedure.

The principle of functional basing

The principle of functional basing is an attempt to deal with the problem of composite support for the denture by registering the mucosal tissues in their functional form rather than in their resting form. Due to the nature of the problem, this can only be accomplished by employing a two step impression procedure with two impression materials of differing physical properties. A preliminary impression in a stock impression tray lacks accuracy on two counts.

First is that the impression material registers the supporting mucosa in its resting form, so that it is not capable of resisting masticatory pressure under occlusal loading at the moment when support is needed. In order to overcome this, the mucosa must be captured in its functioning form, so that the base can be related to the metal framework in the same relationship that exists between the abutment tooth and the supporting mucosa when an occlusal force is applied to the base

Second, the peripheral borders do not accurately represent the precise extension, into the surrounding soft tissues, that they should have in order to contribute maximally to the support, stability and retention of the prosthesis.

Determining the need for a functional impression

As the displaceability of the mucosal covering of the residual ridge varies within a very wide range from mouth to mouth, it will be found that, as a practical matter, the degree of displacement is so slight in some mouths, with well formed ridges and a thick, healthy mucosal covering that the functional impression is not essential. This is particularly apt to be true in the maxillary arch.

The contour of some mucosal foundations may be virtually the same under an applied load as it is rest. Accordingly, the decision as to whether or not to employ the functional impression technique in a given case may be based in part on the following test.

Add acrylic resin bases to the frame work by forming the resin base directly on the stone cast after a tinfoil substitute has been applied to it. place the frame work, with the attached resin bases, in the mouth and apply pressure with the fingers to each base. If the base can be depressed a discernible amount with finger pressure, a functional impression procedure should be carried out. If, however, there is no discernible movement of the base, consideration may be given to dispensing with the functional impression technique. When such a decision has been made, the resin bases can be used as record bases to obtain the required intraoral records.

Impression materials for the functional impression

Several impression materials are available for registering the functional impression, any one of which can be relied upon for recording accurately the mucosal tissues in their supporting form, provided they are properly manipulated.

The objective of the functional impression is twofold:

(1) To register the mucosa in its supporting form and

(2) To establish precisely accurate functional borders for the denture

Several methods may be employed for establishing accurate peripheral borders and registering the mucosa in its supporting form[80]. Two well known ones are the *altered cast* method cast method and *the functional reline* method. The altered cast method is carried out before the denture bases are processed. The functional reline method is accomplished after the denture bases have been processed onto the framework.

THE ALTERED CAST METHOD

The altered cast method is composed of the following steps:

1. An acrylic resin impression base is added to the retention latticework of the framework. It should resemble, in outline the approximate area to be covered by the finished denture base. The mandibular base should cover the retromolar pad and extend onto the buccal shelf. The maxillary base should cover the hamular notches and the palate.

2. The framework is taken into the mouth and the attached impression bases are adjusted so that they are approximately 2 mm. short of their estimated functional length. Clearance should be provided for any muscle attachments of frenula, which are impinged upon by the tray borders by trimming the acrylic resin to accommodate them. Softened modeling composition is applied to the borders of the impression base, one section at a time, and the border tissues are gently manipulated to create a rounded contour that extends to the flexure line at the mucobuccal fold.

3. The functional impression may be registered with low fusing wax, with zinc oxide eugenol paste, or with one of the rubber base impression materials. The procedure mentioned here is that using impression wax.

4. The tissue surface of resin tray is thoroughly dried, and a uniform layer of number 4 mouth temperature wax is applied with a small brush. The frame work should be seated in the mouth for approximately 5 minutes, being certain that all parts of the metal that contact tooth surfaces are precisely and accurately in position on the teeth.

5. The framework is removed from the mouth and the impression is dried and inspected. Areas of the wax that are in good functional contact will appear glossy, while areas of insufficient contact will appear dull. Wrinkled areas indicate a lack of firm contact or insufficient time for the wax to flow. The impression is complete when the entire tissue surface has a glossy appearance and the peripheries are rounded smooth.

Altering the cast

The master cast is altered by substituting the denture bearing areas, obtained with the corrective impression material, for the ones that were registered in the first impression. *Ansari* suggested the use of separating wax during pouring the cast for easy separation of edentulous ridge areas of the cast.

The altered cast procedure is commonly used for *bilateral distal extension* mandibular removable partial dentures to promote optimal adaptation of the extension bases[59]. The stability of a distal extension removable partial denture may be enhanced by the use of an altered cast impression technique. In the altered cast procedure, the denture framework is fabricated on a one-piece cast and used as the base for a relieved individual tray, which is then used to make a second impression of the edentulous tissues. Since the metal framework can be stabilized on the teeth while the impression is made[71], the mucosal tissues are in as nearly a state of rest as possible. The second impression is used to alter the cast in order to reproduce in the new registration, the relationship between the edentulous and the dentulous regions. Then, the altered cast is used to form the basal surfaces of the denture bases.[71]

Chen et al[14] described an altered cast impression technique that eliminates conventional past dissecting and impression boxing. However, *Sykora*[84] pointed out that the same technique was described by *James*. Only the discussion how to box the impression was not touched upon by James.[40,41] Chen stated that the technique is without problem. However, *Sykora* suggested that a minute movement along the line of rotation of the removable partial denture frame may occur while the new full-arch impression tray is inserted and superimposed with irreversible hydrocolloid over the remaining teeth, framework and distal extension acrylic resin tray. This will alter the relationship between the distal extension base, the frame and remaining teeth. It will be undetected at this stage and will only become apparent at the try in appointment with the discrepancy in the inter-arch space.

Principle

Basically, the altered cast procedure applies some of the principles of impression for complete denture to the fabrication of the tissue surfaces of extension removable partial dentures.[52] The refined impression of the edentulous tissue surface is made after the metal casting has been completed and is used to alter the edentulous areas of the master cast.

Advantages

1. Altered cast impression best fulfills the principle of controlled tissue support.

2. The use of altered cast technique during the preparation of distal extension removable partial dentures highly recommended because it is easy to perform and gives predictable results. It is common for distal extension removable partial denture to rock antero-posteriorly if altered cast impression has not been loosed.

3. Remarkable stability, positive occlusion, decreased stress, decreased adjustment required with this technique.

4. It provides a more favorable support

5. Altered cast technique cause least displacement of tissue.

Disadvantages

1. It is time consuming and require a specific visit

2. It requires difficult laboratory phase

3. The occlusal record cannot normally be made at the same visit at which the altered cast impression is made.

4. Errors may be introduced in orienting and stabilizing the framework on the sectioned cast and additional appointment is necessary for making inter-occlusal records.

THE FUNCTIONAL RELINE METHOD

The denture is fabricated in the customary manner, with the sole exception; that a layer of relief metal approximately 1 mm in thickness is burnished to the residual ridge areas of the processing cast prior to final closure of the flask. The purpose of the metal is to provide a space on the tissue side of the denture for the reline material. The needed space could be provided by grinding the side of the denture with stones and burs, but since the metal ensures a uniform depth, it is the preferred method.

The denture is processed and fitted into the mouth in the customary manner, except that the relief metal is left in place. When it has been worn for a trial period of a week or 10 days, and all needed adjustments have been accomplished, the relief metal is stripped from the acrylic resin and a reline procedure is carried out. The peripheries are refined by border molding with modeling composition, after which the final impression is obtained with one of the corrective materials, fluid wax, one of the rubber base impression materials or zinc oxide eugenol paste.

When the denture is inserted, with the impression material inside it, due care should be exercised to ensure that the framework is in its proper position on the teeth. For this reason, the open mouth technique is preferred, because it permits observation of the framework on the teeth at all times. When the impression has been obtained, a cast is poured and the reline processed, either by flasking or by means of reline jig.

Disadvantage

A disadvantage of the reline method is that the occlusion, which has been so carefully developed, may be altered slightly by the reline impression and may require adjustment following the processing of the acrylic resin.

Another drawback is the possibility of creating a visible junction line between newly added acrylic resin and the original base material

SECTIONAL IMPRESSION

Sectional impressions are used for altering the master cast or relining finished denture bases. *Leupold* [50] compared bases adapted to master cast poured in impression made in full arch peripherally moulded custom trays and bases adapted to altered cast produced by the same material in sectional trays attached to metal frame works. They found that the sectional impression trays resulted in less vertical movement of distal extension bases.

El-Sheikh et al [20] stated that the sectional impressions for mandibular distal extension removable partial denture bases have been highly recommended and deemed mandatory to control tissue displacement under the denture bases. Sectional impressions are customarily used to alter the master cast. They confirmed that the sectional impression decreased the vertical displacement of denture bases. The amount of decrease was much less than that reported by *Leupold*.[51] The final impression made in full arch custom trays are clinically comparable to sectional impressions for distal extension mandibular removable partial dentures.

Keng et al [45] described a sectional impression for standing teeth in an otherwise dentulous arch. The procedure uses a 2 part tray that incorporate *zinc oxide eugenol* impression paste and irreversible hydrocolloid material.

Procedure:

On a preliminary cast, fabricate a closely fitting acrylic resin custom impression tray for the edentulous portion of the arch with two positioning projections for the alignment of a second section to allow for relief around the missing tooth. Perforate the tray for retention of material.

Position the edentulous portion of the tray in the mouth and border mould the periphery with green compound and make the maxillary impression. Remove the edentulous portion of the impression and examine it for accuracy. Remove any excess zinc oxide eugenol material around the gingival margins of the remaining teeth with a scalpel blade.

Reseat the zinc oxide eugenol impression in the mouth. Load the cap tray with irreversible hydrocolloid and seat it into position over the tooth. Ensure localization of the tray stops. When set, carefully removed the two part tray assembly as one unit.

The final impression satisfies the requirements of the functional impression for the edentulous ridge together with the necessary detail of the remaining tooth or teeth.

FLUID WAX IMPRESSION TECHNIQUE

The term 'fluid-wax' means only that the impression wax is added in the fluid form by painting it on in subsequent additions.

Making the impression waxes

To avoid overheating, but more especially to maintain a uniform temperature of the impression wax, at which it can be manipulated best and in the least possible time, melting of the wax should be done in a double boiler. The waxes can be added most smoothly, if they are kept at about 200° to 210 °F.

Preliminaries to the correctable impression

- If the metal unit of the partial denture is to carry the fluid wax registration, it is advisable that the appliance be worn for several hours previous to the making of the impression. Usually, it is placed with a temporary base and allowed to remain until the following day and then the impression is completed.

- Frequent removal and insertion of the appliance is necessary in making a correctable impression. It is essential that this be done easily. If there is any interference encountered after the initial period of wear, the areas should be located and relieved.

- Inspect the surface structures in the area to be covered by the base. There must be no area of abrasion, hyperemia, edema or other abnormal condition of the mucosal pad.

- After being in the mouth, the surface of the temporary base should be freed of any retained mature alba or greasy residue.

- Some space should be made on the under surface of the temporary base.

- A supply of Korecta-wax should be maintained constantly at a temperature just below the boiling point of water.

- The brush which has been found satisfactory for applying the wax is one with medium still bristles about 3/8 inch in them.

Positioning the correctable wax impression

The keying of an impression of an edentulous area of supporting ridge to the remaining teeth must be done with exactness is that both the ridge and the abutment support simultaneously. To properly seat an appliance, which is carrying an impression, apply pressure in three places with fingers of the right hand. The first finger should press in the area of the occlusal rest and minor connector, the second finger rests on the linguo plate indirect retainer, while the third finger contacts the occlusal rest of the left bicuspid area in the mandibular class-I situation.

The minimal pressure to be used is that amount needed to seat the appliance firmly, if, however, a pressure is exerted which approximates the patient's biting force, then an additional advantage is gained. Under the latter load, the abutment periodontal membrane is given the same tensile load that it will receive in function. Functional loading of the teeth should be continued throughout the making of the correctable impression.

Making the correctable impression

There are three rather well defined divisions in the clinical procedures of obtaining a satisfactory registration of the subjacent structures by the correctable wax impression method.

1. Extend the area of base coverage

2. Achieve placement of the surface areas which yield

3. Provide adequate release of any surface against which too much impression material may have been placed.

Maximum coverage means that the peripheries have been extended to the line of the most explained movements of the circumjacent structures. To determine the line of maximum movement, the most extreme actions should be simulated. The appliance must be kept completely seated as these exercises take place. If no wax turn shows at the periphery, add an excess of wax just inside the margin and repeat the action.

If considerable extension is needed, stronger reinforcement may be obtained by heating a portion of a small paper clip and letting it melt into the outside surface of the extended area. All addition of reinforcing wax should be chilled before another correction of the impression is attempted. The reinforcement must stop short of the periphery so that the excess impression wax is allowed to escape and so that it is turned at the correct flange length. When the proper length of a border is attained, its thickness is determined by adding impression wax until the outside surface becomes glossed. The impression wax becomes glossy in areas, which are being supported. An area is dull when it is not receiving support. A thin film of impression wax is repeatedly added to the dull areas only.

The procedure should be one of building up to a relation of support, rather than, reducing a surplus wax down to a relation, which is free from over displacement of the underlying structures. Each time an addition of impression wax is made, the appliance is reseated completely to position against the teeth for about one minute before exercise of border structures is begun. Compressed air must be used to dry a surface thoroughly before adding more wax.

If the wax or resin base or the metal structure should be so close to the surface as to show through the impression wax, remove some has to provide a greater thickness of wax. Be sure that there is ample thickness of the impression wax in the area adjacent to an abutment so as to avoid any tissue displacement.

After these objectives are attained, a corrective wax impression will have been developed to this stage. All functional impression do not require the same period of time for third phase of the procedure to be satisfactorily accomplished. It will depend upon the size of the impression, how far the excess wax must be moved in order to reach the periphery and be molded by the moving border structures. To be better able to evaluate the time needed, following factors involved in its determination should be considered-

- The rate of flow of the impression wax being used

- The temperature of the wax when it reaches contact

- The pressure being exerted on the wax

- The degree to which the wax is free to escape

- The distance which must be traveled by the excess material

- The amount of excess wax to be moved

- The time during pressure is continuously maintained.

It should be policy, however, to leave the correctable impression in position for several minutes as the last step in the impression procedure. It is better that it be retained too long than that it be removed too soon.

The various requirements for acceptable completion are

- All of the contacting surface is glossy

- The roll or turn at the wax periphery is continuous, except where the tissue does not move during functional activity

- The appliance has been kept under continuous seating pressure for a long enough period to let any excess wax escape to the periphery

- Any of the base wax or framework, which was showing through the impression wax, has been relieved.

- The periphery has been thickened to record the bulk, which the cul-de-sac structures can accommodate.

- The impression has been removed from the mouth without having the periphery distorted by coming in contact with tongue or cheek or anything, which could deform it.

The fluid wax impression is so readily corrected, that this type of impression makes it possible to determine very accurately the line of peripheral movement characteristic of each patient. Unlike some methods of impression making, the variables can be so well controlled in fluid wax impression procedure that exceptionally few adjustments are needed after insertion of an extension base appliance. This approach can be employed on most dentures with a distal extension denture base on one side and a tooth supported denture base on other side.

Applegate's cold wax impression technique and subsequent modifications of it by various authors make the impression of dentulous and edentulous regions in separate steps. The results obtained are good but the procedures are difficult.

An objection made to the correctable wax impression procedure has been that the appointment time required for it is excessive. However, *Applegate* stated that the correctable impression fluid wax technique for registration of the subjacent structures, when correctly executed, can be defended as a time saving device, rather than being regarded as a method which is excessively time consuming.

Modified wax impression procedure

Preiskel [63] described a modified wax impression procedure. When at last six and preferably more, teeth remain in a dental arch with a reasonably prominent curvature, another of Applegate's technique can be used. The abutment preparations are completed and the master impression is poured.

An acrylic resin tray is constructed on the master cast. They tray covers the two edentulous ridges and a rigid U-shaped strut of acrylic resin is carried over the occlusal surfaces of the remaining teeth and preparations. This strut contacts the teeth and three widely spaced points, usually at the midline and at the most distal abutment preparations on both sides. The handle of the tray is joined to the strut. The temporary crowns are removed and the tray is placed in the mouth to ensure that the distal extension bases are correct and that the three occlusal stops make proper contact.

The impression surface of the tray is painted with wax and the tray is firmly positioned in the mouth with the occlusal stops in place. When the impression has been made, the edentulous ridges are cut off the cast, and the impression tray is reoriented back on the cast by means of the occlusal stops. The master cast obtained in this manner provides an accurate representation of the mucosa of the residual ridges under slight load in its correct relationship to the abutment preparations and to the remainder of the dental arch. Where only six anterior teeth remain, another method must be used. The metal work of the anterior crown and of the removable prosthesis is cast, and the attachments are soldered in position. Acrylic resin bases are attached to the removable prosthesis, the crowns are positioned in the mouth, and an impression with selective mucosal displacement is made with no hinge action or vertical movement within the attachment. The partial denture framework and abutments are removed in their correct relation by means of an alginate impression that is made over both structures.[50]

FUNCTIONAL IMPRESSION

Rapuano[64] described a functional impression technique for making distal extension removable partial denture based philosophically on Hindel's load distribution principle that uses a dual impression procedure in a single tray

An irreversible hydrocolloid impression was made that included the mandibular teeth and residual ridges. A custom resin tray was constructed on this cast. The teeth and adjacent tissue on the diagnostic cast were relieved with four thickness of asbestos paper. A resin-separating medium was applied to the cast and a cold curing rein was adapted over the relieved teeth and residual ridges. The tray was tried in the patient's mouth to verify the accuracy of its fit. The borders were sufficiently reduced to allow space for border molding. The part of the tray covering the teeth was removed completely. Approximately 2 to 3 mm of space existed between the internal surface of the tray and the teeth. The borders of the tray were molded in the edentulous region and then an impression of the residual ridges was made using a zinc oxide eugenol impression paste.

The maxillary and mandibular posterior wax planes were placed in the patient's mouth and the mandible was guided into the centric relation record[69]. The occlusion rims were in intimate contact over the total length of the residual ridges. The patient was instructed to close forcefully on the occlusion rims. If the closing force was strong, the rims should have been made of a resistant material such as modeling compound.

The impression of the teeth was made with an irreversible hydrocolloid impression material. The impression material was loaded into a plaster gun, and the teeth were painted through the large opening in the tray with the impression material. Then, the impression material in the plaster gun was injected through the same opening. The impression material was forced to flow around the teeth and through the multiple perforation in the tray. Once the injection was completed, the border molding was completed after the impression material had set, the impression was removed.

The same technique is excellent for an internal attachment distal extension removable partial denture. Some the advantages of this procedure are.

1. The denture bases are functionally loaded by the patient as he closes his jaws in centric relation at the proper vertical dimension of occlusion. This force stimulates the functional load that will be placed on the finished removable partial dentures[73].

2. The functional load is applied along the entire length of the posterior occlusion of the removable partial denture.

3. A second try is not needed

4. The dual material single tray produces a less bulky final impression from which a master cast can be poured.

5. Internal attachment distal extension removable partial dentures have a definite, accurate, functional relationship between the residual ridges and the abutment crown when this procedure is used. [7]

STATIC RELINE TECHNIQUE

Steffel[80] described static reline impression technique. For this impression, very little preparation or modification of the tissue side of the basal seats is required. These surfaces must nearly be out of contact with tissues when the stabilizing points are in contact with the teeth and should be roughened.

An impression paste (ZnOE) is applied in limited abundance, and the partial denture is carried to the mouth with the fingers pressing against the anterior framework and the thumbs under the chin. The points of contact are held firmly against the natural teeth. While this is being done, no attention is directed to the distal extension. The paste type of static impression literally makes itself as it fills the spaces and flows between the slightly upraised bases and the undisturbed ridge tissues. The resulting imprint records the tissues without distortion.

When the impression paste has hardened, the denture is removed, the fin like excess are eliminated and the reline impression literally makes itself as it fills the spaces and flows between the slightly upraised bases and the undisturbed ridge tissues. The resulting imprint records the tissues without distortion. If the partial denture is an all-metal type, the reline material is replaced by plastic, which is mechanically locked to the base, or by a metal casting of the same proportions.

When the reline denture, now ready for service again, is seated back to place, the metal frame work will be accurately reoriented with reference to the natural teeth at the same time as there occurs intimate relationship of the basal seats to the ridge tissues. The occlusion however, will be heavy, the degree of occlusal disparity depending upon the distance the extensions had raised when pressure was applied anteriorly in the impression making. This interfering occlusion is now reduced only to the level of anticipated tolerance and the teeth coordinated with their opponents by grinding all correcting being done on the lower artificial teeth if the denture is opposed by natural one, on both upper and lower teeth if both dentures are artificial[86].

FUNCTIONAL RELINING IMPRESSIONS

If the partial denture has only the two clasps with no indirect retainer(s), an effort should be made to refine the occlusion before going forward with the relining impressions. If there is an indirect retainer, or additional points of contact of framework to make tripoding or stabilization possible, it is preferred to re-harmonize the positional relationship of the partial denture to teeth and tissues first and to correct the occlusion first also, before proceeding with the refine impression.

After the occlusion has been corrected with the denture being supported at its proper relative elevation, the impression substance, undercuts are eliminated from the base, and some base materials.

An excess amount of low fusing modeling composition is applied to the bases excess is used so that there will always be some material present wherever needed. The appliance is mass heated in the water bath and inserted part way toward the tissues but not seated. The patient is not allowed to close for contact. Usually about four mass beatings and insertions progressively, are necessary before allowing the patient to close natural teeth to light centric occlusal contacts. For final closure, the patient should be instructed to proceed just to point of faint facet or interdigitative contact, not asked to close with great force. Doing so might result in an upward mandibular displacement

The border limits are now formed by heating short sections with the torch then manipulating the cheeks and lips for border outline. Impression substance, and even base material if necessary, is now removed all along the crest of the ridge, and usually along the mylohyoid ridge. Also, if the distal extensions are long escape holes are bored completely through the base material at these areas requiring relief.

Impression wax is painted over the entire impression surface, the partial denture is inserted, and the patient instructed to gradually force it to place, then to continue with the functional movements of chewing. The wax must always be well chilled before removal for examination. After a short period in the mouth, excess of wax should be removed and the restoration reinserted. The patient is again instructed to go through the functional movements of chewing. Limited movement of the base extension under the functional stresses will mold the soft wax for relief of vulnerable areas. Final step in the reline impression does not change the previously, re established position of the partial denture in the arch, but merely refines the impression and specializes the fit with reference to the different types of supporting tissues.

Bollwerk [8] described an improved impression technique for partial dentures with distal extension saddles. In this technique, an impression of the entire arch is made in either an alginate or reversible hydrocolloid material and poured in stone. A clear acrylic tray is formed over this cast. The tray is fitted to the dental base area and the borders are trimmed in the same manner as for a full denture impression. They are trimmed two millimeters short of the reflection of the buccal and enough on the lingual to prevent the tray from lifting when the tongue is fully extended out of the mouth.

The outline form of the denture base area is trimmed with Kerr green-stick modeling compound to include the retromolar pad, the masseters plane and buccal pouch. The lingual roll is developed by the proper tongue movement. When a satisfactory peripheral outline has been developed on the both side and the anterior lingual area, some of the compound is trimmed form the area of the lingual frenum. This area will be recorded in a zinc oxide eugenol wash impression.

The occlusion rim is built up on the saddle area of the acrylic tray. The occlusal surfaces of the occlusion rim are softened and the stock tray used in the final impression is placed over this to apply pressure on the soften surface. To facilitate the removal of both impressions in one piece, horizontal grooves are cut in the occlusion rim on the lingual and the buccal sides to create a mechanical lock between the two materials. After the trays and the occlusion rim have been developed, the handle of the acrylic tray is cut off, and enough acrylic is cut away to expose fully the lingual surfaces of all of the teeth.

An impression of the tissue bearing area is made in a zinc oxide-eugenol impression material. The mouth is dried with sponges prior to making this impression. After this impression has set, the impression material is removed from the inter-dental spaces and the lingual surfaces of the teeth without removing the impression from the mouth.

A water-jacketed tray filled with reversible hydrocolloid is placed over the occlusion rim, try and zinc oxide eugenol impression. Pressure is exerted by this stock tray against the occlusion rim. The amount of pressure is determined by the type of soft tissue on the ridge area. The hydrocolloid impression material is chilled by water. After five minutes, the hydrocolloid material will be set and both impressions are removed from the mouth in one piece

HINDEL'S TECHNIQUE

An acrylic resin tray is processed on a cast made for an impression that should include all areas of future tissue support of the partial denture. The tray is selectively relieved and when checked in mouth, should cover the edentulous areas upto the border tissue attachments and should include the retromolar pads. The bases of the tray should be connected with each other by means of an acrylic resin lingual bar. The bar should cover the area between the muscle attachments of the floor of the mouth and the lingual gingiva of the anterior teeth. The tray should clear the free gingiva around the abutment teeth.

The tray is loaded with an easy flowing zinc oxide eugenol paste and is brought into position in the mouth. After the material has hardened, the tray is removed and the impression examined. The material that has flowed from the tray on to the abutment teeth should now be cut away.

The next step is to make an impression of the teeth and to establish a relationship between the teeth and the mucosa in a displaced state. For this purpose, a perforated tray that has been provided with two circular openings of approximately 18 mm diameter in the region of the first molars is used. The impression of the soft tissue areas is placed in the mouth.

Then, while the tray is being loaded with an irreversible hydrocolloid impression material, some of this material is used to fill the space between the soft tissue impression and the remaining teeth. The loaded metal tray is then inserted over both the teeth and the acrylic resin tray. The index fingers are pressed through the openings in the perforated tray until they contact the underline tray, then pressure is exerted on it. this pressure should be maintained until the alginate impression material has set[36]

The completed impression is that remove as one unit. The cast made in this impression will be a reproduction of both the surface of the teeth and the undistorted surface of the mucosa, but the two will be related to each other with the mucosa in a functional state.

IMPRESSION PROCEDURES FOR FIXED PARTIAL DENTURES

Missing teeth can be replaced with fixed prosthesis that will improve patient comfort and masticatory ability, maintain the teeth and integrity of the dental arches and in many instances, elevate the patient's self image.

An accurate set of full mouth working casts are a necessary part in the construction of fixed restorations. To obtain the working casts, the materials used are plaster, alginate and modeling compound. The impression materials most frequently used for cast restorations are elastic when removed from the mouth. The combination mixture of light and heavier – bodied materials may be used in a disposable stock tray or in a custom made acrylic resin tray for making impression for fixed partial dentures.[29]

The special syringe is used to inject the light bodied material about the crown preparations as a preliminary to positioning a tray with the heavier bodied material over the prepared teeth.

Individual impression of abutments may be made in copper bands.[46] These impressions are made one or two at a time and not removed from the abutments until an impression in the same material of the entire jaw is made over the individuals in the bands. Upon removal from the mouth and individual impression will come away with the complete arch impression and maintain the correct relationship between the teeth, tissue and abutments. Better subgingival coverage is possible when banded impression are used and thin fins of inter proximal tissue are eliminated

The research of *Skinner* and *Cooper, Jorgensen, Rosensteil,* Thiokol Chemical Corporation and Farihurt confirms the belief that the synthetic rubber base is desirable impression materials, which possess qualities equal to that of the hydrocolloids.[22] The laboratory procedures for pouring the impression and separating casts are also equally important.

INITIAL IMPRESSION

Impression material

Diagnostic cast are made using alginate impression material, and a preliminary treatment plan is developed.[74] A tray of auto polymerizing acrylic resin is constructed on the diagnostic cast.

Tray selection

Select a stock impression tray, which will allow equal distribution of impression material when inserted in the mouth. Adjust the trays by bending the sides to ensure that the retromolar pad and the maxillary tuberosity are covered by the tray. The tray is tried in the mouth and adjusted to avoid impingement on the soft tissues. With the mandibular tray, instruct the patient to touch the tip of the tongue to the palate, this will facilitate seating of the tray. Check the vestibular extension of the tray by reflecting the lips and cheeks.

Impression procedure

A chemically accurate cast can be made from good irreversible hydrocolloid impressions. Making irreversible hydrocolloid impression is an important preliminary for adequate diagnosis. Use of irreversible hydrocolloid impression material requires a disciplined step-by-step approach.

Prior to making the maxillary impression, have the patient rinse with mouthwash, and wipe the oral cavity free of excess saliva with a gauze pad.

Pour the measured water into the mixing bowl; and add the irreversible hydrocolloid. Spatulate the mixture thoroughly, to eliminate bubbles and attain a smooth creamy mixture.

Place the mixture in the impression tray, and wipe a small amount of irreversible hydrocolloid material onto the occlusal surfaces of the teeth.

Seat the irreversible hydrocolloid mixture in the tray for a full arch impression. The impression should be held rigidly in position until after the alginate sets. Since the mouth tissues are at a higher temperature than that of the tray, the alginate material will start of gel first in the region adjacent to the tissues. After two minutes, remove the impression by breaking the posterior seal with a snapping motion to prevent tearing of thin areas of the impression material and distortion. Remove excess irreversible hydrocolloid from the patient's mouth and instruct the patient to rinse with mouthwash.

Rinse the impression with a gently stream of cool tap water to remove any saliva or small particles; inspect the impression for adequate extension and detail. When a satisfactory impression is achieved, wrap it in a moist paper towel to prevent shrinkage. Position the impression to prevent any pooling of water in the occlusal surfaces. Pour the impression immediately in gypsum product to form the diagnostic cast.

Reversible hydrocolloid impression material has several significant advantages compared to the polysulfide and silicone elastomeric materials. It is less technique sensitive, and its set is not affected by slight moisture in the gingival sulcus.

Skinner and Carlisle[76] concluded that the accuracy of the impressions with the alginate impression materials was slightly better than that obtained with the agar materials. According to the investigators, the principal objection to the use of alginate impression material for indirect restorative procedures is that the gelation time of product was too short. The gelation time of an alginate impression can be retarded chemically with better results than if the retardation is effected by the used of cold water for mixing. However, the chemical retardation should be effected by the manufacturer, and not by the dentist.

Hoblit[75] showed that there was no difference between reversible (agar) and irreversible (alginate) impression material so far as the accuracy of impression obtained was concerned.

CUSTOM TRAY FABRICATION

Two layers of base plate wax are adapted over the teeth of diagnostic casts, extending at least a tooth beyond the teeth to be included in the impression[9]. *Calomeni*[12] suggested that the tray is made using 1.3 to 4 mm spacer of wet asbestos or wax on the cast instead of the usual 2 mm spacer. The extra space in the tray makes it possible to cut escape ways for the flow of injection rubber-base material used in making the final impression.[19]

Stops are provided on the occlusal surfaces of teeth not involved in the restorations. The unprepared tooth or soft tissue acts as an anterior and posterior stop in positioning the tray during impression making. The cast and the wax shim should have separating medium applied prior to placement of the acrylic a resin. The tray is made with quick-setting acrylic resin material. A 2mm clearance between inner surface of the tray and the teeth is required in the areas of preparation.[65]

The tray should extend beyond the cervical margins of all teeth. The tray must be rigid and thick enough to prevent distortion or bending and have a strong handle to facilitate removal of the impression from the mouth. Sharp or rough edges are removed from the periphery of the tray. Silicon adhesive is applied lightly to all surfaces and permitted to dry.[60]

TISSUE DISPLACEMENT

Obtaining a complete impression is complicated when some or all of the preparation finish line lies at or apical to the crest of the free gingival. In these situations, the preparation finish line must be temporarily exposed to insure reproduction of the entire preparation. Control of fluids in the sulcus is also necessary, because liquids can cause an incomplete impression of the critical finish line area. A variety of gingival displacement techniques has been described in the literature. The methods have included one or more of three techniques: mechanical, chemico-mechanical and surgical.

1. **Mechanical**: Physically displacing the gingiva was one of the first methods use for insuring adequate reproduction of the preparation finish line.[42] The most common technique is to place retraction cord in the gingival sulcus after the teeth have been prepared. A copper band or tube can serve as a means of carrying the impression material as well as a mechanism for displacing the gingiva to insure that the gingival finish line is captured in the impression.[21]

   One end of the tube is festooned or trimmed to follow the profile of the gingival finish line. The tube is filled with modeling compound and seated carefully in place along the path of insertion of the tooth preparation. The use of copper band can cause incisional injuries of gingival tissues. But recession following their use is minimal ranging from 0.1 mm in healthy adolescents to 0.3 mm in general clinical population [16]. Copper bands are especially useful for situations in which several teeth have been prepared. The use of copper band could negate the necessary of remaking the entire full arch impression just to capture one or two preparations.

   A rubber dam also can accomplish the exposure of the finish line needed. Generally it is used when a limited number of teeth in one quadrant are being restored and in situations in which preparations do not have to be extended very far subgingivally.

   Plain cotton cord was used for sulcus enlargement physically pushing away the gingiva from the finish line. However its effectiveness is limited, because the use of pressure alone will not control sulcular hemorrhage.

2. **Chemico-mechanical:** By combining chemical action with pressure placing enlargement of the gingival sulcus as well as control of fluids seeping from the walls

of the gingival sulcus is more readily accomplished. Caustic chemicals such as sulfuric acid, trichloracetic acid and zinc chloride have been tried. But their undesirable effects on the gingiva led to their abandonment.

Over the years, racemic epinephrine has emerged as the most popular chemical for gingival retraction. Surveys document that cord impregnated with 8% racemic epinephrine is the most commonly used means of producing gingival retraction.

However, there is controversy surrounding the use of epinephrine for gingival retraction. Epinephrine causes an elevation of blood pressure and increase in heart rate. Some investigators have found that the physiologic changes that occur when epinephrine impregnated cord is placed in an intact gingival sulcus are minimal.

Placing the retraction cord

The operating area must be dry and evacuating device is paced in the mouth and the quadrant containing the prepared tooth is isolated with cotton rolls. The retraction cord is drawn from the dispenser, bottle with sterile cotton pliers and a piece of approximately 5.0 cm long is cut off. It a twisted or wound cord is used, grasp the ends between the thumb and forefinger of each hand. Hold the cord taught and twist the ends to produce a tightly wound cord of small diameter. If a braided or woven cord is used, twisting is not necessary.

The retraction cord should be moistened by dipping it 25% aluminum chloride solution in a dappen dish. Cord impregnated with either epinephrine or aluminum sulphate are twice as effective when saturated with aluminum chloride solution prior to insertion into the gingival crevice. If there is slight hemorrhage in the gingival crevice, it can be control by the use of haemostatic groups.

From the cord into a 'U' and loop it around the prepared tooth. Hold the cord between the thumb and forefinger and apply slight tension in an apical direction. Gently slip the cord between the tooth and the gingiva in the mesial interproximal area with a Fischer packing instrument or DE plastic instrument IPPA. Once the cord has been tucked in on the mesial, use the instrument to lightly secure it in the distal interproximal area.

Proceed to the lingual surface and begin working from the mesiolingual corner around to the distolingual corner. The tip of the instrument should be inclined slightly toward and area where the cord has already been placed.

Continue on around to the mesial, firmly securing the ord where it was lightly tacked before. Cut off the length of cord protruding from the mesial sulcus as closely as possible to the interdental papilla. Continue packing the cord around the facial surface, overlapping the cord in the mesial interproximal area. The overlap must always occur in proximal area, where the bulk of tissue will tolerate the extra bulk of cord.

Pack all but the last 2-3 mm of cord. This tag is left protruding so that it can be grasped for easy removal.

3. **Surgery:** The surgical techniques can be further broker down into rotary curettage and electro surgery

   a. *Rotary curettage:* Rotary curettage is a 'troughing' technique, the purpose of which is to produce limited removal of epithelial tissue in the sulcus while a chamfer finish line is being created in tooth structure. The technique, which also has been called 'gingettage', is used with the subgingival placement of restoration margins.

   The removal of epithelium from the sulcus by rotary curettage is accomplished with little detectable trauma to soft tissue, although there is a lessened tactile sense for the dentist. Rotary curettage, however, must be done only on healthy inflammation-free tissue to avoid the tissue shrinkage that occurs when diseased tissue heals.

   Suitability of gingiva for the use of this method is determined by three factors; absence of bleeding upon probing, sulcus depth less than 3.0 mm, and presence of adequate keratinized gingiva.

   In conjunction with axial reduction, a shoulder finish line is prepared at the level of the gingival crest with flat end tapered diamond. Then a torpedo diamond of 150-180 grit is used to extend the finish line apically one half to two thirds, the depth of the sulcus, converting the finish line to the chamfer. A generous water spray is used while preparing the finish line and curetting the adjacent gingiva. Cord impregnated with aluminum chloride or alum is gently placed to control hemorrhage. The cord is remove after 4 to 8 minutes, and the sulcus is thoroughly irrigated with water. This technique is well suited for use with reversible hydrocolloid.

Several studies have been done to compare both the efficacy and the wound healing of rotary curettage with those of conventional techniques. *Kamassky* et al reported less change in gingival height with rotary curettage than with lateral gingival displacement using retraction cord. With curettage, there was an apparent disruption of the apical sulcular and attachment epithelium, resulting in apical repositioning and an increase in sulcus depth. The changes were quite small, however, they were not regarded as clinically significant.

Tupac and *Neacy*[72] found no significant histological differences between retraction cord and rotary curettage, *Ingraham* et al reported slight differences in healing among rotary curettage, pressure packing and electro surgery at different time intervals after the tooth preparation and impression. However, complete healing hand occurred by three weeks

b. *Electrosurgery:* There are situations in which it may not be feasible for desirable to manage the gingiva with retraction cord alone. The use of electro surgery has been recommended for enlargement of the gingival sulcus and control of hemorrhage to facilitate impression making. It has been described for the removal of irritated tissue that has proliferated over preparation finish lines and it is commonly for that purpose. An electrosurgery unit is a high frequency oscillator or radio transmitter that uses either a vacuum tube or a transistor deliver a high frequency electrical current of at least 1.0 Mhz.

Electrosurgery produced controlled tissue destruction to achieve a surgical result. Current flows from a small cutting electrode that produces a high current density and a rapid temperature rise at its point of contact with the tissue. The cells directly adjacent to the electrode are destroyed by this temperature increase.

There are different forms of current that can be generated for electrosurgical use.

i. Unrectified, damped current characterized by recurring packs of power that rapidly diminish.

ii. Partially rectified, damped current that produces a waveform with a damping in the second half of each cycle.

iii. Fully rectified, filtered current that is a continuous wave that produce excellent cutting.

For the patent's safety, it is important that the circuit be completed by the use of ground electrode. The safe use of electrosurgery dictates that current flow be facilitated along the proper circuit from the generatory to the active electrode, the patient and back to the generator. The proper grounding of a patient is considered to be the single most important safety factor.

Electrosurgery should not be employed on patients with cardiac pace makers. It can produce sparks in use, so it should not be used in presence of flammable agents.

Technique

Before an electrosurgical procedure is done, verify that anesthesia is profound an reinforce it if necessary. With a cotton tipped applicator, place a drop of a pleasant smelling aromatic oil at the vermilion border of the upper lip. The odor from it will help to mask some of the unpleasant odor emanating from the mouth during electrosurgery check the equipment to make sure all the connection are solid.

Proper use of electrosurgery requires that the cutting electrode be applied with very light pressure and quick, deft strokes. The pressure required has been described as the same needed to draw a line with an ink-dipped brush without bending the bristles. It is obvious that the electrode is being guided and not pushed through the tissue.

To prevent lateral penetration of heat into the tissues with subsequent injury, the electrode should move at a speed of no less than 7 mm per second.

Initially set the power selector dial at the level recommended by the manufacturer and make adjustments as necessary. As the electrode passes through the tissue, it should do so smoothly without dragging or charring the tissue. If the tip drags and collects shreds of clinging tissue, the unit has been placed on a setting that is too low. On the other hand, if the tissue chars or discolours or if there is sparking, the setting is too high. If an error must be made initially, it is better to have a setting that is slightly too high. Moist tissue will cut best. If it dries out, spray is lightly

A high volume vacuum tip should be kept immediately adjacent to the cutting electrode at all times draw off the unpleasant odors that are generated. Stock frequently to clean any fragments of tissue from the electrode by wiping it with an alcohol soaked sponge

FINAL IMPRESSION PROCEDURE

IMPRESSION MAKING WITH HYDROCOLLOID

As only one accurate cast can be made from a hydrocolloid impression, two impressions are made; a sectional (quadrant) impression for making a die and full arch impression for the working cast. Be sure that the patient has adequate anesthesia. If the impression is being made at a separate appointment subsequent to the preparation of the tooth, if is necessary to anesthetize the area again. Select the tray to be used and try it in the mouth to make certain that if fits.

Place adhesive plastic strips into the tray to keep the teeth from pushing all the way through to the tray when it is seated in the mouth. Position two of the stops, one on top of the other, at the rear of each side and in the front of the tray on full arch trays. On quadrant trays, place stops at the front and rear. Make sure that the stops will contact unprepared teeth.

Isolate the quadrant containing the prepared tooth, insert the retraction cord, and place a gauze pack in the mouth. Fill the impression tray with a tube taken from the storage bath and place the filled tray in the tempering bath.

Screw a short, blunt 19-guage needle onto the end of the anesthetic type syringe. Remove a plastic, anesthetic type cartridge containing hydrocolloid from the storage bath. Insert it into the syringe and express a small amount of material to make sure it is flowing freely.

Remove the gauze square from the patient's mouth. Blow a very light water/air mist on the prepared tooth before removing the retraction cord. The cord should be slightly moist, but not wet. Do not blow compressed air on the tooth after the retraction cord has been removed.

Carefully remove the cord from the sulcus by grasping the free end in the mesial interproximal area with a pair of cotton pliers. Tease the cord out gently so that hemorrhage will not start. If an impression is being made of multiple preparations, remove the cord from the sulcus around each tooth, one at a time, immediately before the material is injected. Inject hydrocolloid from the syringe into the sulcus, starting in an interproximal area first. Hold the tip above the mouth of the crevice, taking care not to drag the tip along the gingiva. Proceed smoothly around the entire circumference of the preparation, pushing impression material before the tip.

Kishimoto[47] described the disposable syringes developed for reversible hydrocolloid. Their main component is a plastic cartridge, similar to the anesthetic cartridge, similar to the anesthetic cartridge. It is used in a non-aspirating anesthetic syringe with a blunt, half-inch long, 19-gauge disposable needle and was first described for use with reversible hydrocolloid in 1937.

It was observed that when used with polyether or polysulfide, some disposable syringes were found to require in excess of 100 pounds of force to operate. The metal reusable syringe required only 13 and 21 pound respectively. With the same material, the air-activated syringe required none.

There was no significant difference among the syringes with regard to bubble entrapment during impression making. The air activated syringe was most easily controlled, as evidenced by the significantly smaller number of tracing errors made with it.

There were highly significantly fewer bubbles in impression material extrusions from the second half of the syringe full than from the first half. The smaller tip orifices 0.6 mm and 0.67 mm in diameter caused highly significantly fewer bubbles in the extruded impression materials than did the larger tips.

The pneumatically powered syringe provided an easier means of extruding the impression material through small syringe tip orifices than did the conventional hand powered syringe.

An alternative way of applying the syringe material is called the 'wet field technique'. The prepared teeth are bathed in warm water, and syringe material is deposited in generous quantities only on the occlusal surfaces of the teeth. The relatively viscous tray material is counted on to force the lighter bodied syringe material on to the sulcus as the tray is seated. This technique should be used only on preparations that do not contain internal features, such as grooves boxes, or isthmuses.

Nichols[62] described a simple technique to obtain sharper, more complete sub gingival detail in difficult subgingival preparations. It consists of painting in mild detergent on the prepared teeth immediately before making the impression. The detergent used is liquid soap diluted one part to three with tap water. Upon completion of tooth preparation, the soaked pellet is rubbed onto the prepared tooth and all of the other teeth in the arch while the assistant directs a combination air and water spray onto the pellet.

This pellet scrub and spray is used to remove the debris that has accumulated on the teeth during the preparation. Just before making the impression, the retraction threads are removed and the region is flushed with water and dried with air. The brush is dipped into the detergent solution and the solution is brushed into the gingival sulcus and completely over the prepared tooth.

The detergent should be left wet on the teeth but the excess should be carefully removed. It acts as lubricant, preventing thin edges of the hydrocolloid from adhering to the inside of the sulcular wall and tearing on removal of the impression. When the gelled impression tray is removed from the mouth, the sulcular detergent application allows the hydrocolloid material to slide away easily without tearing.

The assistant should remove the sectional tray from the tempering bath, wipe the surfaces of the hydrocolloid free of water, and connect the tray to the hoses. Give the syringe to the assistant in exchange for the tray. Seat the tray while the assistant connects the hoses to the unit. Hold the tray in place for 8 minutes while cool water flows through its tubes. Do not have the patient hold the hydrocolloid tray. It is too unstable, and a distorted impression could result.

While the sectional tray is setting in the patient's mouth, the assistant can fill the full arch tray and place it in the tempering bath. Remove the sectional impression with a quick motion, along the long axis of the teeth. Check it for completeness, and wash it in cold tap water. Remove the excess moisture from the surface of the impression with air, but do not desiccate the material. Spray the impression or place it in an appropriate solution to disinfect it before pouring it.

Below a very light water/air mist on the preparation and inject hydrocolloid around it again. It is usually not necessary to repack the sulcus with cord for the working cast impression. Place the full arch tray and connect the tubes to the unit. Hold it in place for 6 minutes. Remove it with a snap.

Electroplating the impression

Preparing the impression

After removal from the mouth, the impression is rinsed with running cold water to remove saliva and is then dried with air. Using a brush, silver powder is burnished onto the areas to be plated. The excess powder is blown away using an air syringe. Any areas that are not covered with silver powder are retouched with the brush and the excess powder is blown away. The silver powder is applied to the teeth that have been prepared for restorations and also to the teeth immediately adjacent to them. If it is desired to have the saddle area of edentulous spaces reproduced in silver, these areas are also covered with silver powder[39]. The powder is continuous form on tooth to the next and goes onto the adjacent gingival and alveolar regions.

Attaching the connecting wire

The connecting wires carry the plating current to the silvered surface of the impression. Two or more wires should be strategically located in areas that have been silvered. The wires should be located far enough away from preparation and other critical areas. The wires are so placed as to carry current to the sides oppose to the area to be plated to ensure rapid and even plating. The wires are also conveniently placed to serve as a handle to support the impression in the bath.

The insulation is removed from both ends of each wire, 1 inch of wire being exposed at one end and 1/8 inch at the other end of each wire. The shorter exposed end of each wire is cut diagonally to from a sharp point. The shaper end of the wire is inserted into the rubber base impression until the insulation meets the surface of the impression and no bare wire is exposed. The other wires are similarly placed, and the wires are brought together and twisted around each other until the bare ends form one unit.

Placing the impression in the Bath

The impression is immersed in the solution of the plating bath, and the connecting wires are attached to the cathode terminal. A rubber bulb pipette filled with plating solution is used to wash away the trapped air. The current is then switched on and the rheostat is adjusted to give a current equivalent to 10 mA for each tooth in the impression to be plated. Deposition of silver begins at once and can be recognized by a change in the color from metallic grey of the silver powder to a yellowish grey. Plating of the impression to continue for short time until a thin flash of silver covers the surface of the impression. Usually from 15 to 30 minutes is required to obtain an even overall flash of silver.

Isolating the Dies

The impression is removed from the plating bath into a small glass dish. The dish is carried to the sink and the impression is thoroughly washed with running cold water. The impression is dried with air and examined carefully to determine if all the parts to be plated are covered with the silver flash

The prepared teeth are isolated with a ring of wax. Electrical continuity is maintained by the thin flash of silver beneath the wax, and the teeth continue to plate

Completing The Plating

The impression is returned to the plating bath, connected as previously, and any trapped air is washed away. The current is set at a value equivalent to 10 milliamps for each tooth to be plated. At these settings, some 12 to 15 hours are required to deposit an adequate thickness of silver.

When the required period has elapsed, the impression is removed from the bath, put at once into a glass dish, thoroughly washed with water and dried with air. The connecting wires are pulled our of the impression and discarded

IMPRESSION MAKING WITH RUBBER BASE IMPRESSION MATERIAL

Custom resin trays have been utilized in elastomeric impression techniques because these material are more accurate in uniform thin layers of 2-3 mm. many authors advise against the use of stock trays because the uneven bulk of impression material is conducive to distortion. However, it has been reported that the mean difference in material thickness between custom and stock trays is less than 1 mm and that variation from uniform thickness exist in both custom and stock trays.

Ortensi [72] described a procedure for fabricating modified custom tray with the acrylic auto polymerizing resin for more predictable and accurate impression making. In this procedure, obtain a model for custom tray fabrication. Apply a layer of wax on soft and hard tissue to create the space for impression material. Create the stock to stabilize tray positioning Fabricate tray by applying a uniform layer of resin. During the appointment for impression making, reline the tray with auto polymerizing resin. Mix the resin with minimum concentration of monomer. Pour the mixture into the tray and wait until it becomes opaque. Place the tray on the teeth and apply a light pressure until it settles completely. When the resin becomes rubbery, remove the tray. Cure the resin in hot water.

Use a fine tipped permanent market to mark abutment contours obtained after relining. After relining, use a multiblade spherical bur to drill 2 mm deep grooves inside the abutments. Join the grooves without getting any deeper into the abutments. For teeth with supragingival finish lines, trim away 2 mm of resin around the margins. For teeth with subgingival finish lines, remove only 0.5 mm of resin from around the edges. In pontic regions, remove 2 mm of resin apply and adhesive on the tray, wait until it dries and proceed with impression making.

Yoder and Thayer [93] described a basic procedure for the use of rubber base impression material for fixed partial denture impressions. The previously formed acrylic impression tray is observed in the month and any undesirable extension or impingement upon the soft tissue is relieved. The tray is dried thoroughly, and a thin coat of tray adhesive is applied to the inside of the tray. The entire surface of the tray to be covered by impression material, including the borders, is coated with the adhesive.

The adhesive is allowed to dry for at least 10 minutes. The drying may be hastened by a few blasts of air. The double mix technique is used. A light bodied rubbed-base impression material is used for syringe application to the tooth preparation, and a heavy bodied material is used for filling the tray. Both of these materials are measured and mixed according to the manufacturer's directions. The accuracy of the die made from rubber base impression materials depends on the bulk impression material used.

The impression material is mixed to a homogenous consistency by either spatulation or stirring[79]. The blade of the spatula is covered with the brown paste. By drawing the brown into the white paste, it will be easier to mix the material and clean the spatula later. The mixing time is approximately 45 seconds for most rubber base impression materials.

The syringe material is mixed and aspirated directly from the pad or from a dappen dish into the barrel of the syringe. The tray material is mixed in a similar manner as the syringe material, and the tray is loaded by picking up the entire mass of the mix on the spatula

All teeth and the surrounding tissues must be thoroughly dry. The filled syringe is carried into the mouth and the tip of the syringe nozzle is placed into the deepest part of the most distal crevice. The impression material is injected into the crevice and around the circumference of the tooth by continuous pressure on the syringe until the abutment is completely covered. The tip of the nozzle of the syringe is kept in contact with the tooth and buried in the rubber base impression material at all times[70]. All small amount of syringe material is ejected onto the teeth and under the lip in the opposite dental arch to serve as a guide to the progress of the setting of the material. The loaded tray is carried into place, seated with a firm direct pressure, and then merely stabilized to prevent its movement while the impression material sets.

After the impression material has set completely, the tray is removed in such a way as to minimize distortion and prevent tearing. The impression is rinsed and dried gently so that the inner surface may be examined for correctness.

IMPRESSION MAKING WITH POLYSULFIDE

Be sure that the patient has adequate anesthesia. Try the custom tray in the mouth to make sure it fits without impinging on the prepared tooth. Insert the retraction cord and place a large gauze pack in the mouth.

The following steps require an assistant. On one disposal, mixing pad squeeze out 1.5 inches (4.0 cm) each of light (syringe) base and accelerator. On a second pad place, 5.0 inch (12.5 cm) strips of regular (tray) base and acceleration. Pull the plunger from the injection syringe and set it aside.

The assistant should start mixing the tray material on one pad 30 seconds before the operator begins mixing the syringe material on the other. Pick up the dark accelerator on the spatula and incorporate it into the white base. Holding the spatula flat against the pad, mix with a back and forth motion, pressing hard against the pad. Change directions often to produce a smooth, homogenous mixture. Be careful not to incorporate bubbles. Do not take more than 1 minute to mix it.

Fold a sheet previously removed from the mixing pad in half and then fold it to make a cone. Open it up and wipe the syringe material from the spatula onto the crease. Fold the cone over. Squeeze the syringe material from the cone into the back end of the syringe. Insert the plunger and express all the air from the syringe.

In a second method of loading the syringe, the back end of the syringe is brought in contact with the pad, and quick, closely spaced sweeps of the syringe will fill it, with a minimum of material spilled. In a third method, the syringe tip is removed and the front end of the syringe is buried in a collected mass of material on the pad or in a dappen dish.

Remove the gauze spares from the patient's mouth. Be sure that the retraction cord is slightly damp before removing it from the sulcus. Immediately inject polysulfide syringe material into the sulcus. Hold the tip just above the mouth of the crevice. Do not drag the tip along the gingiva. Proceed smoothly around the entire circumference of the preparation, pushing impression material ahead of the tip. Continue around the preparation until the entire tooth is covered.

Use an air syringe to direct a stream of air against the material to spread it evenly over the surface of the preparation and drive it into small details. Impression material is also forced more completely into the gingival crevice. Excessive pressure, prolonged air application, and use on patients with a thin band of attached gingiva should be avoided because of possibility of producing interstitial emphysema. Seat the tray slowly until the stops hold the tray solidly in one position. The tray should be held with light pressure for 8 to 10 minutes without movement. The set of the material can be tested with a blunt instrument. When the material rebounds completely without leaving any trace, it has set.

After the material has polymerized, the impression is removed. The wings on the sides of the tray can be used for added leverage in this task. Rinse the impression to remove blood and saliva. Blow it dry and inspect it. an impression of the opposing arch can be made with alginate. Soak the impression in an appropriate disinfectant solution before pouring it.

IMPRESSION MAKING WITH CONDENSATION SILICONE

Before the preparation is begun, select a stock tray that fits the arch. Coat the inside of the tray with a thin, even coat of adhesive and allow it to dry. For a full arch impression tray, place two scoops of putty (base) on the pad. Use one scoop for a sectional tray. Add six drops of accelerator for each scoop of base. Incorporate them on the pad with a spatula for a few seconds. Then transfer the material to the palm of the hand and kneaded it for 30 second. The material should be streak free.

Roll the base into a cigar shape and place it into a stock impression tray. Cover the base with a polyethylene spacer, and seat the tray in the mouth. Remove the tray from the mouth after the initial set has occurred. Peel off the spacer and remove any excess on the periphery of the tray with a sharp knife. Set the tray aside for use after the tooth has been prepared. Be sure that the patient has adequate anesthesia. Isolate the quadrant containing the prepared tooth, place the retraction cord, and insert a gauze pad in the mouth. The following steps require an assistant. Squeeze out 8 inches (20 cm) of the thin wash silicone base onto the disposable mixing pad. Use 4 inches (10 cm) for a sectional tray. Add one drop of accelerator per inch of base. Mix with a spatula for 30 seconds, the mix should be free of streaks. Place about one-third of the wash material into the back end of the syringe. While you are inserting the plunger and expressing air, the assistant should place the rest of the material into the tray.

Remove the gauze square from the patient's mouth. Be sure that the retraction cord is slightly damp before removing it from the sulcus

Carefully remove the cord from the sulcus by grasping the free end in the interproximal region with cotton pliers. Tease the cord out gently so that hemorrhage will not start. Immediately inject syringe material into the sulcus. Hold the tip just above the mouth of the crevice. Do not drag the tip along the gingiva. Proceed smoothly around the entire circumference of the preparation, pushing impression material ahead of the tip. Do not skip any areas, but continue around the preparation until the entire tooth is covered. Give the syringe to the assistant in exchange for the loaded tray.

Seat the tray slowly until it is firmly in place. They tray should be held in place with no downward pressure for 6 minutes.

After the material has set, remove the impression as quickly and as straightly, as possible. Rinse the impression to remove blood and saliva. Blow it dry and inspect it. Soak the impression in an appropriate disinfectant solution before pouring it.

One of the impression technique developed for use with silicon base impression materials is the putty wash or relining technique. The technique obviates the need for a custom tray because the preliminary impression made with stock tray is essentially a custom made tray formed by the putty material.

Then carried out a study concerned with the adhesion of light bodied silicone to a putty silicone in a putty wash impression technique when the preliminary putty impression was contaminated with human saliva and residue from acrylic resin used in fabrications of provisional restorations by direct technique.

Data indicated that condensation and addition silicones differed in their susceptibility towards the tested contaminates. Salivary contamination and chemical residues from the autopolymerizing acrylic resins weakened the bond strength and caused adhesive failure depending upon the type of silicone impression material used.

The most significant disadvantage of silicone rubber as a dental impression material is the high degree of shrinkage. To overcome this advantage, a moderately filled material has been developed for the single mix impression technique; however, the control of shrinkage is limited because too much filler results in the material too heavy to be used in a syringe. Another means of overcoming the disadvantage is the use of a combination of highly filled material and a lightly filled material in the double impression technique. This technique, however, involves two operations. *Fuasyma* et al developed a new technique called the laminated single impression technique. In this technique, the heavy and wash type are mixed at the same time. The wash type is laminated in a thin layer on the surface of the heavy type, and this is loaded in a tray and immediately impressed upon the preparation. The purpose of this lamination is to prevent the direct contact of the heavy type with the preparation. The body of the heavy type also drives the wash type into the gingival sulci without use of a syringe.

Tjan et al [84] showed that the condensation silicone produced under sized dies and there was virtually no difference in shrinkage and distortion in the delayed pours.

IMPRESSION MAKING WITH POLYVINYL SILOXANE

Paint the custom tray with adhesive at least 15 minutes before the impression is to be made. If a tube-dispensed material is used in a double mix technique, the assistant and operator start mixing material at about the same time. Mix with a spatula for about 45 seconds until all steaks are eliminated. Then load the syringe and tray.

If a cartridge system is used, load a cartridge of light bodied material into one dispense and cartridge of medium or heavy bodied material into another. Remove the gauze pack placed earlier in the patient's mouth. Be sure that the retraction cord is slightly damp before removing it from the sulcus. Carefully remove the cord and inject the impression material, starting in one interproximal area and pushing the material ahead of the tip. While the dentist applies the light bodied material with a syringe, the assistant loads the tray with the medium or heavy bodied material. Exchange the syringe for the loaded tray and seat it firmly in the mouth. Hold it in place for 7 minutes from the start of mixing.

Remove the impression as quickly and straightly as possible to prevent distortion. Rinse, blow it dry, and inspect it. place it in a disinfectant solution before pouring it.

IMPRESSION MAKING WITH POLYETHER

It is imperative that the operator be well organized and execute swiftly while using this impression material coat the custom tray with the adhesive supplied with polyether. Express approximately 7.5 inches (19cm) each of base and accelerator onto a disposable mixing pad. Mix with a spatula for about 1 minute until all streaks have been removed. Load the back end of the syringe. The material sets too fast and is too viscous to use a paper cone. The assistant should load the tray while the operator proceeds. Remove the gauze pack from the patient's mouth, which was placed earlier.

Be sure that the retraction cord is slightly damp before removing it from sulcus. Carefully remove the cord from the sulcus and inject the impression material, quickly but carefully, starting in one interproximal area. Exchange the syringe for the loaded tray and seat the tray firmly in place in the mouth. Hold the tray in place for 4 minutes. Remove the impression. Rinse it and blow it dry. Inspect the impression and treat it with disinfectant.

DUAL ARCH IMPRESSION PROCEDURE FOR F.P.D.

Dual arch impression technique represents, a significant advance in fixed prosthodontics and have many advantages over conventional impression techniques.[44]

Types of dual arch trays

The types of dual arch trays can be categorized as follows; metal or plastic, with sidewalls or "sideless" and by the amount and location of the arch enclosed by the tray.

According to the amount had location of the arch enclosed by the tray, they are categorized in posterior sextant, quadrant, three quarters of an arch, anterior sextant and full arch.

Selection of the tray

The first step consists of selecting a tray that fits passively and does not impinge on any of the teeth or on any anatomical structure, such as the alveolar ridge. The tray should not interfere in any way with occlusion. If it does not meet these requirements, the dentist should try a tray of different shape or size or use a conventional impression technique.

Orienting the impression tray

After selecting the appropriate tray, the dentist should practice inserting the tray into the exact position required to produce an adequate impression. The tray should capture as much of the adjacent and opposing teeth as needed to produce an adequate impression and occlusal registration. Distinguishing features on the tray can be used to establish a repeatable orientation that facilitates insertion to the proper position.

Technique

1. ***The one step, dual arch impression technique*** is similar to conventional impression techniques except that the opposing teeth are impressed at the same time that the prepared tooth and adjacent teeth are impressed. In the one step technique in which the plastic tray is used, the dentist injects low and medium viscosity impression material around the prepared tooth. Bite registration material or putty is inserted on both sides of tray, which is then inserted into position, and the patient closes into maximum intercuspation.

2. ***There are mainly two, two-step dual arch impression techniques;*** the most popular ones are hydraulic pressure technique ad laminar impression technique. Both techniques require a preoperative impression before the tooth is prepared. This can be accomplished with a metal or plastic tray.

    a. *Hydraulic pressure technique:* Low of medium viscosity impression material is injected around the prepared tooth and into the preoperative impression of the unprepared tooth. The dentist reinserts the preoperative impression and the patient closes into maximum intercuspation. The generated hydraulic pressure forces the wash material into the sulcus and around the preparation. The dentist should take care not to inject too much impression material around the prepared tooth or into the preoperative impression of the prepared tooth. Vent holes can be drilled on the buccal or lingual aspect of the impression material and tray to allow the escape of excess wash material

    b. *Laminar impression technique:* In this technique, the dentist takes the preoperative impression in bite registration material. The preoperative impression of the tooth to be prepared is relieved to the depth of 0.5 mm in the cervical area. The dentist then drills two holes through the buccal wall of the impression tray and impression material. One hole is on the mesial aspect of the prepared tooth and the other is on the distal aspect.

The tray is reinserted and medium or low viscosity impression material is injected through the mesial hole. The material flows around the prepared tooth and into the sulcus, eventually, it exists through the distal hole. This technique produces an accurate impression of the prepared tooth within the preoperative impression.

Advantages

1. Dual arch impression technique enables the dentist to capture an impression of the prepared tooth, the opposing teeth and the occlusal registration in one procedure. This saves chair time

2. Dual arch impression technique requires far less impression material than conventional impression procedures.

3. Dural arch impression technique can yield castings that require little occlusal adjustment. Accuracy is excellent.

4. They can be used comfortably by the dentists working unassisted or with an assistant who has limited experience taking impression

5. The laminar impression technique is specially effective in areas that are difficult to isolate

6. The one step dual arch impression technique has the advantage of requiring less time than the two step techniques and is more familiar to dentists and assistants.

Disadvantages

These procedures are technique sensitive and can produce poor results when improperly performed

DOUBLE ARCH IMPRESSION FOR F.P.D

The double arch impression technique is a mature concept that has been used effectively. [10]

Impression tray

Double arch impression trays can be used with any type of impression material. To establish a registration of complete closure of the teeth in centric occlusion, thin, disposable inserts of paper or a thin gauge mesh can be used.

A disposable impression tray (triple tray) is intended to facilitate the making of double arch impressions. Evaluators stated that the size, strength and comparative low cost of the triple tray are improvements over some other double arch trays.

Two clinical variables can be controlled with double arch trays. *First, the physical deformation of the mandible* during opening is eliminated. *Second, the natural shifting of the teeth* to assume a maximum inter digitations can be registered with this method.

Several variations in the double arch impression technique are possible and tray selection is dependent on the demand of each patient.

The decision to use the double arch impression technique must be based on sound physical principle. The success of any impression is partially dependent on selection of correct tray for a particular clinical situation. In the double arch technique, this is specially essential because it is a quadrant technique and centric occlusion must be recorded correctly.

Technique

Wilson and Werrin[91] described the double arch impression procedure. Before loading the impression tray, it should be tested in the patient's mouth. If the tray has a distal crossbar, it should be positioned posterior to the last tooth. The patient is instructed to close the jaws several times to check the reproduction of the centric occlusion. Complete inter-digitation of the tooth should be confirmed on both sides of the dental arch.

Tissue treatment of the gingival sulcus prior to impression making can be managed by electrosurgery or the use of retraction cords or rings. When the soft tissue is ready, the impression material is mixed and loaded on both sides of the tray.

Although polysulfide's can be used satisfactorily, stiff bodied impression materials are preferred. A polyether or an addition reaction silicone impression material assures adequate tray rigidity, postpone imminent dimensional change, and exhibits low permanent deformation.

Immediately after the removal of retraction cords or rings, the prepared teeth the sulcus spaces are injected with a light bodied impression material. The loaded tray is positioned carefully in the mouth, and the patient is instructed to close into centric occlusion.

When the impression material has set, the impression is removed in a single movement with an equal bilateral pressure on the flanges of the tray. The impression, counter impression and occlusal registration have been produced simultaneously. The excess impression material is trimmed from the buccal and lingual sides of the tray with scissors, and the counter side is poured first with either artificial stone. Next, the working side is poured with small increment of stone or resin. The impression of the prepared teeth should be slightly overfilled before a dowel pin is placed in it.

Vibrations

1. Under certain conditions, copper tube impression can be used effectively in double arch trays. The tooth preparation must be short occluso-gingivally to obtain a satisfactory tube impression and still allow the tube to be trimmed to clear the opposing teeth in centric occlusion. Provision must be made for simultaneous withdrawal of the tube with the impression in the double arch tray.

2. Making an initial impression to be used as custom tray for a localized impression of a prepared tooth can be accomplished with double arch trays provided the initial impression can be reseated easily into the proper centric occlusion.

3. A conventional method of providing temporary restorations is possible if the double arch impression is made before the tooth preparations are begun. After the preparations are finished, the teeth are lubricated with a separating medium to minimize pulpal responses. The preparation side of the impression is filled with a self-curing resin formulated for

provisional coverage and carefully reseated in the mouth into centric occlusion. When the temporary restoration has set, it can be removed from the impression. The impression can then be used when needed as a custom tray to make a wash impression of the tooth preparation.

Advantages

1. The technique is uncomplicated. It reduces chair time operating costs and the possibility of errors.

2. The tray permits complete closure in centric relation and recording of an undistorted occlusal registration. Minimal occlusal adjustment are inherent in double arch impression technique

3. Patient and dentist expense, time and efforts are saved because the double arch technique requires fewer steps

4. Triple tray is more comfortable for the patients than many other designs. Patient gagging is reduced, because it is a quadrant rather than a full arch tray.

5. Disposable tray eliminate cleaning, inventory expense and the possibility of disease transmission from one patient to another.

COMBINATION IMPRESSION FOR F.P.D

Hypothetical benefits of combined reversible hydrocolloid irreversible hydrocolloid impression system were noted as early as 1951.[4] The irreversible tray type hydrocolloid serves to unite the reversible material to the impression tray. Prepared with water, the cool temperature of this surrounding irreversible hydrocolloid causes the underlying injectable reversible hydrocolloid to solidify. Since there is no tray type reversible hydrocolloid used with this system, the equipment for it is not needed.

Dentloid, an injectable reversible hydrocolloid bonds to irreversible hydrocolloid to form a combined reversible hydrocolloid/irreversible hydrocolloid impression system[3]. The reversible hydrocolloid is prepared and injected around the preparation. An impression tray with the irreversible hydrocolloid is placed over the preparations, surrounding structure and the impression material. The two impression materials unite permitting removal of the compound impression. A study indicated that the dimensional stability for combinations of Dentloid with Jeltrate is within accepted clinical standards.

Feinberg[23] described combination impression as follows: Use modified tray for alginate, which covers the entire jaw. Prepare the modified trays by cutting out part of the occlusal portions of the regular alginate trays. The cutout section includes the region distal to the canine tooth posteriorly to the end of the tray, except a small strip of metal left at the distal end of the tray to hold it together the trays with these cutout areas are used for posterior restorations. For anterior restorations, the occlusal portion of the anterior segment is removed.

Cover the cutout parts of the tray with a layer of red wax, and lute it in place on the tray. Removable metal covering may be used for this purpose but are not necessary fill the tray with a mix of alginate impression material and make an overall impression of the entire jaw. Try to obtain an impression of as much of the mouth as possible, including the reflected border tissue.

Remove the alginate impression from the mouth and cut away part of the alginate impression material in the region where the restorations are being made and remove the red wax, which is over this section. To make provision for plaster, cut the alginate from the mesial side of the anterior abutment to the distal of the posterior abutment. Cut away enough alginate lingually to provide space for the necessary thickness of plaster to hold the copings in place. Remove all of the alginate from the buccal side in this region. For a lower impression, all of the alginate is removed from the lingual side of the teeth so that the plaster will be held by the buccal and lingual sides of the tray.

Replace the prepared alginate impression back in the mouth. It will snap into its correct position if there is a complete impression of the tissue up to the reflected border tissues. Make a thin mix of quick setting plaster. With a small spatula, place the plaster through the hole in the occlusal surface of the tray so it surrounds the copings and covers the edentulous areas of the ridge. Place an excess of plaster in the opening in the tray and condense it by pressing the piece of red wax into place over the opening in the tray. Allow the plaster to set and then remove the impression from the mouth.

Brotman[10] described combination impression for fixed restoration. It will permit a relatively simple full arch impression to be made by using plaster of Paris and alginate impression material in the same tray at the same time. In this procedure, the copings prepared are placed in position on prepared teeth. A rim lock tray which fits comfortably in the mouth is selected. The center of the tray is placed at the midline of the arch, and the external surface of the tray is marked with a pencil at the position of the most anterior abutment tooth. The tray is removed and set aside to receive the impression material.

Two persons are required for the preparation of the impression material. The alginate impression material is mixed by the dentist and plaster of Paris is mixed by the assistant, both mixes completed at the same time. The plaster is placed carefully in the edentulous and abutment areas in the tray and allowed to extend slightly beyond the abutment position as indicted by the pencil mark. The alginate material is then placed in the remainder of the tray and brought into direct contact with the plaster. The completely filled tray is then placed in position in the mouth and both materials are allowed to set at the same time. When both materials are completely set, the entire tray is removed.

IMPRESSION AND SIMULTANEOUS BITE REGISTRATION

Mack[54] described the technique for accurate impression and bite in crown and bridge. The technique uses a rigid quadrant impression tray that enables the operator to take the master impression, the counter impression and the bite all in one operation.[27]

Mark a linked vertical line that is in line with the midline of the maxillary central incisors. The opposite side should be marked so that the inked lines will still be visible while the impression is in the mouth. The inked line should be blown dry and patient instructed to open and close to check the accuracy of the line.

In the clinical technique, injection rubber is used in syringe and heavy body rubber in the tray. It involves taking a no. 4 brush bending the brush and cutting off a third or a half of the bristle length so that the bristles will be rigid enough to push the injection rubber around the gingival margin of the preparation. Just before injecting rubber around the preparation, a small amount of rubber is injected on the tray cover.

During the impression, the injection rubber is exuded around the gingival margin upto only half the length of the preparation. Then the bent brush is dipped into the puddle of rubber on the tray cover and rubber is continuously pushed in one direction around the preparation, gently being pushed down into the sulcus.

Then the syringe is used again to pour out more rubber around the whole preparation and the occlusal surfaces of both jaws, after which the two sided tray containing heavy body rubber is inserted in mouth in mouth and patient is instructed to bite down. The bite should be examined by checking the inked lines.

This technique simplifies the entire procedure as it necessitates only one impression ensure an accurate bite and eliminates the need for a wax bite.

Scott et al[70] described the check bite impression using irreversible alginate/reversible hydrocolloid combination to obtain the working cast, opposing cat and the articulation in one clean, economical and time saving procedure. The procedure is described as follows:

Insert a wettable paper to separate the impressions of two arches. When the prepared teeth are ready for impression making and the tissue retraction is complete, load the hydrocolloid carpule into the syringe and the irreversible hydrocolloid to both sides of the impression tray. Carefully injected the prepared tooth and adjacent structures together with the opposing occlusal surfaces.

Position the loaded tray and ask the patient to close to the desired occlusal position and maintain tooth contact for 3-4 minutes. Remove the tray from patient's mouth and then pour the impression side in a compatible die stone and place the dowels for the prepared teeth at least one adjacent tooth. Allow the stone to set

IMPRESSION PROCEDURE FOR PIN RETAINED RESTORATION

To make an impression of the preparation for a pin retained restoration, nylon bristles must be used to duplicate the pin holes. Impression materials will not fill the small diameter holes being used.[56]

If a kit is being utilized, use the nylon bristle supplied with a given diameter drill. It will be approximately 0.002 inch smaller in diameter than the drill. If it is necessary, shorten the bristle to prevent it from hitting the impression tray and being distorted. Cut it with a sharp scalpel.

Place a bristle in each of the pinholes. Proceed with the impression in the usual manner, making sure to inject all the way around the head of the bristle. Withdraw the impression in the line of draw of the preparation and pins. Pulling the impression off in another line may tear the bristle out. Pour the impression in the usual way. When the stone has set, separate the impression and cast.

In general, methods for making impressions of teeth prepared for pin-retained restorations have involved the introduction of a standard pin into each pin canal. The technique will produce casts with pin canals whose diameter is smaller than that of the tooth canals. If the diameter of the standard pins were not smaller than that of the tooth canals, the pins would stick in the pin canals when the impression is removed.

Mollersten[61] developed an impression technique in which casts can be made exactly reproducing the pin canals. In this technique, a smooth cylindrical metal pin with a slightly smaller diameter than the twist drill is used for the pinholes and a light bodied Thiokol rubber is used for impression. The diameter of the metal pin must not be smaller than that of the pin going into the gold casting.

Pontopins or Inca steel pins may be used. They exist for this purpose in three dimensions, their diameter being 0.75, 0.65 and 0.55 mm. respectively and correspond to the 0.8, 0.7 and 0.6 mm drills. The impression surface is thoroughly dried. The walls of the pinholes are covered with a thin layer of petroleum jelly or oil. The pins are covered with the light bodied Thiokol rubber and placed in the pin canals. An individual tray is made of modeling compound supported by a plastic or stock metal tray. This carries the heavy bodied Thiokol rubber to place for the completed impression. The adhesion between the two rubber compounds is sufficient even if the rubber material in the pin canals should harden before the final impression is made.

This technique offers following advantages.

- The pins will adhere to the canals, thus eliminating the risk that the pins will be displaced when the final impression is made

- Since each pin is completely covered with Thiokol rubber, the adhesion of the pins to the impression is improved

- Removing the pin from the die is easier since the pin is completely covered with the elastic Thiokol rubber.

- This method enables the dentist to use pins that are nearly the same diameter as the pin holes

- There is no risk of deformation of the pins, since they are metal

Goransson[30] described a Pontostructor method of using vertical parallel pins in firmed partial dentures combined with a new impression technique for pin ledge preparations. This technique involves the use of reversible hydrocolloid impression combined with special plastic pins.

Prefabricated plastic pins are available in three different sizes. These are 0.78 mm for 0.8 mm pin hole, 0.68 mm for 0.7mm pin hole and 0.58 mm for 0.6 mm pin hole. The diameter of the pin is 0.02 mm less than the prepared pin hole in the tooth. The pins are 5 and 6 mm in length with a head that is 1 mm in diameter. The head of the pin ensures that the pin will be retained when the impression is removed from mouth. The plastic pins are inserted into the pin holes in the prepared teeth by means of special tweezers which hold the pin firmly. Reversible hydrocolloid impression material is used. The pins are firmly retained in the hydrocolloid impression and can be easily withdrawn from the tooth. The impression must be removed in the direction of the overall long axis of prepared teeth. Artificial stone is vibrated into the impression to form the cast.

BIBLIOGRAPHY

1. **Akerly WB**. A combination impression and occlusal registration technique for extension-base removable partial dentures. J Prosthet Dent 1978;39:226-9

2. **Appegate OC.** The partial denture base. J Prosthet Dent 1955; 5:636-49

3. **Appleby DC, Cohen SR, Racowsky LP, Mingledorff EB.** The combined reversible hydrocolloid / irreversible hydrocolloid impression system: Clinical application. J Prosthet Dent 1981;46:48-58

4. **Appleby DC, Pameijer CH, Boffa J.** The combined reversible hydrocolloid/irreversible hydrocolloid impression system. J Prosthet Dent 1980;44:27-35

5. **Banuman R, DeBoer J.** A modification of the altered cast technique. J Prosthet Dent 1982;47:212-3.

6. **Blatterfein L, Klein IE, Miglino JC.** A loading impression technique for semiprecision and precision removable partial dentures. J Prosthet Dent 1980;43:9-14.

7. **Blatterfein L.** The use of the semiprecision rest in removable partial dentures. J Prosthet Dent 1969;22:307-32.

8. **Bollwerk EH.** An improved impression technique for partial dentures with distal extension saddles. J Prosthet Dent 1953;3:476-80.

9. **Bomberg TJ, Hatch RA, Hoffman W.** Impression material thickness in stock and custom tray J Prosthet Dent 1985;54:170-2.

10. **Brotman IN.** Combination impression for fixed restorations. J Prosthet Dent 1955;5:667-9.

11. **Buckley GA.** Diagnostic factors in a choice of impression materials and methods. J Prosthet Dent 1955;5:149-61.

12. **Calomeni AA.** A wash technique using rubber base impression material. J Prosthet Dent1971;25:520-5.

13. **Charbeneau GT.** Principles and practice of operative dentistry, 3rd ed. Varghese Publishing House, Bombay, 1989. p. 374

14. **Chen S, Eichhold WA, Chien CC, Curtis S.** An altered cast impression technique that eliminates conventional cast dissecting and impression boxing. J Prosthet Dent 1987; 57:471-4.

15. **Civijan S, Huget EF, Simon LB.** Surface characteristics of alginate impressions. J Prosthet Dent 1972;28:373-8.

16. **Coelho DH, Brisman AS.** Gingival recession with modeling plastic copper band impressions. J Prosthet Dent1974;31:647-50.

17. **De Van MM.** Basic principles in impression making. J Prosthet Dent 1952;2:26-35.

18. **Dumbrigue HB, Esquivel JF.** Selective pressure single impression procedure for tooth mucosa supported removable partial dentures J Prosthet Dent 1998;80:259-61.

19. **Eames WB, Sieweke JC, Wallace SW, Rogers LB.** Elastomeric impression materials: Effect of bulk on accuracy. J Prosthet Dent 1979;41:304-7.

20. **El Sheikh HA, Abdel Hakim AM.** Sectional impression fro mandibular distal extension removable partial dentures, J Prosthet Dent 1998;80:216-9.

21. **Ewig S.** Beautiful but glum-porcelain jacket crown. J Prosthet Dent 1954;4:93-102

22. **Fairhurst CW, Furman TC, Schallhorn RV, Kinkpatrick EC, Ryge G.** Elastic properties of rubber base impression material J Prosthet Dent 1956; 6:534-42

23. **Feinberg E.** Technique for master impression in fixed restorations. J Prosthet Dent1955; 5:663-6.

24. **Fenn HRB, Liddelow KP, Gimson AP.** Clinical dental prosthetics, 2nd ed., Staples Press, London, 1961 p. 634

25. **Fuasyama T, Nakazato M.** The designs of stock trays and the retention of irreversible hydrocolloid impressions. J Prosthet Dent 1969; 21:136-42.

26. **Fusayama T, Iwaku M, Daito K, Kurosaki N, Takatsu T.** Accuracy of the laminated single impression technique with silicone materials. J Prosthet Dent 1971;26:146-53.

27. **Getz EH.** Functional "checkbite-impressions" for fixed prosthodontics. J Prosthet Dent 1971;26:146-53

28. **Glossary of prosthodontic terms.** 7th ed. J Prosthet Dent 1997; 81: 48-107.

29. **Going RE.** Accurate rubber base impression. J Prosthet Dent 1968;20:339-14.

30. **Goransson P, Parmlid A.** The pontostructor method and new impression technique. J Prosthet Dent 1965;15:900-7.

31. **Grant AA, Johnson W.** Removable denture prosthodontics. 2nd ed. Churchill Livingstone, Tokyo, 1992 p. 143

32. **Harris WT.** Water temperature and accuracy of alginate impressions. J Prosthet Dent 1969;21:613-7.

33. **Harvey WL.** An improved distal extension removable partial denture base. J Prosthet Dent 1962;12:314-6.

34. **Heartwell CM, Modjeski PJ, Mullins EE, Strader KH.** Comparison of impressions made in perforated and non-perforated rimlock trays. J Prosthet Dent1972;27:494-500.

35. **Hindels GW.** Load distribution in extension saddle partial dentures. J Prosthet Dent 1952; 2: 92-100.

36. **Hindels GW.** Stress analysis in distal extension partial dentures. J Prosthet Dent 1957; 7:197-203.

37. **Hirshberg SM.** Double band rubber impressions. J Prosthet Dent 1965;15:704-9.

38. **Holmes JB.** Influence of impression procedures and occlusal loading on partial dentures movement. J Prosthet Dent 1965;15:474-81.

39. **Hudson WC.** Clinical use of rubber impression materials and electroforming of casts and dies in pure silver. J Prosthet Dent 1958;8:107-14.

40. **James JS.** A simplified alternative to the altered cast impression technique for removable partial dentures. J Prosthet Dent 1985;53:598.

41. **James JS.** A simplified alternative to the altered cast technique for removable partial dentures J Prosthet Dent 1985;53:598.

42. **Jamshidi M.** Technique for accurate rubber base impression for fixed prosthodontics J Prosthet Dent 1982;47:265-8.

43. **Jones JD, Kaiser DA.** A new gingival retraction impression system for a one-stage root form implant. J Prosthet Dent 1998;80:371-3.

44. **Kaplowitz GJ.** Troubleshooting: Dual arch impressions. J Am Dent Assoc 1996;127:234-40.

45. **Keng CB.** Sectional impression for standing teeth in the otherwise edentulous arch. J Prosthet Dent 1996;78:455-6.

46. **Kimmelman BB, Lerman H.** Impressions of single preparations using a copper band shell. J Prosthet Dent 1971;26:154-8.

47. **Kishimoto M, Shillingburg HT, Duncanson MG.** A comparison of six impression syringes. J Prosthet Dent 1980;43:546-51.

48. **Kramer HM.** Impression technique for removable partial denture. J Prosthet Dent 1961;11:84-92

49. **Krammer RV.** A two-stage impression technique for distal extension removable partial dentures. J Prosthet Dent 1988;60:199-201.

50. **Lay LS, Lai WH, Wu CT.** Making the framework try in, altered cast impression, and occlusal registration in one appointment J Prosthet Dent 1996;75:446-8.

51. **Leupold RJ.** A comparative study of impression procedures for distal extension removable partial dentures. J Prosthet Dent 1966;16:708-20.

52. **Leupold RJ, Kratochvil FJ.** An altered cast procedure to improve tissue support for removable partial dentures. J Prosthet Dent 1965;15:672-8.

53. **Mac Gregor AR Fenn, Liddelow and Gimsons's.** Clinical dental prosthetics, 3rd ed., Wright, London, 1989 p. 634

54. **Mack WB.** A new technique for accurate impression and bite in crown and bridge. J Am Dent Assoc 1989;119:297-302.

55. **Mann AW.** A critical appraisal of the hydrocolloid technique: its advantages and disadvantages. J Prosthet Dent 1951;1:733-49.

56. **Mann AW, Courtade GL, Sanell C.** The use of pins in restorative dentistry: I. Parallel pin retention obtained without using paralling devices. J Prosthet Dent 1965;15:502-16.

57. **McGivney CL, Castleberry DJ, McCracken's.** Removable partial prosthodontics, 8th ed, CBS, Publishers and Distributors, India 1989, p. 300

58. **Miller EL, Grasso JE.** Removable partial prosthodontics, 2nd ed. P. 237

59. **Miller TH, Unsicker RL.** Improved method to alter cast in partial denture impression making. J Prosthet Dent1986; 55:135-6

60. **Mitchell JV, Damele JJ.** Influence of tray design upon elastic impression materials J Prosthet Dent 1970; 23:51-7

61. **Mollersten L.** An impression technique for teeth prepared for parallel pins. J Prosthet Dent 1967; 18:579-82

62. **Nichols CF, Woelfel JB.** Improving reversible hydrocolloid impressions of subgingival areas. J Prosthet Dent1987;57:11-14.

63. **Preiskel HW.** Impression techniques for attachment retained distal extension removable partial dentures. J Prosthet Dent 1971;25:620-8.

64. **Rapuano JA.** Single tray dual impression technique for distal extension partial dentures. J Prosthet Dent 1970;24:41-6.

65. **Reisbick MH.** The accuracy of highly filled elastomeric impression materials. J Prosthet Dent1975;33:67-72.

66. **Reisbick MH, Matyas J.** The accuracy of highly filled elastomeric impression materials. J Prosthet Dent 1975;33:67.

67. **Rudd KD, Dunn BW.** Accurate removable partial dentures J Prosthet Dent 1967;18:559-67.

68. **Rudd KD, Morrow RM, Strunk RR.** Accurate alginate impression. J Prosthet Dent 1969;22:294-300.

69. **Santana Penin U, Lozano JG.** An accurate method for occlusal registration and altered cast impression for removable partial dentures during same visit as framework try in. J Prosthet Dent 1998;80:618-8.

70. **Scott GK, Hawkins L, Chetwyn J, Doughty T.** Check bite impression using irreversible alginate/reversible hydrocolloid combination. J Prosthet Dent 1997;77:83-5.

71. **Shifman A.** Index to reposition the metal framework accurately on altered cast, J Prosthet Dent 1991;68:79-81.

72. **Shillingburg Jr HT, Hobo S, Whitsett LD, Jacobi R, Brackett SE.** Fundamental of fixed prosthodontics, 3rd ed. Quintessence Publishing Co, Inc, USA 1997 p. 281

73. **Singer F.** Functional impressions and accurate interocclusal records for removable partial dentures. J Prosthet Dent 1962;12:536-41.

74. **Skinner EW, Hoblit NE.** Study of accuracy of hydrocolloid impressions. J Prosthet Dent 1956;6:80.

75. **Skinner EW, Hoblit NE.** The study of the accuracy of hydrocolloid impressions. J Prosthet Dent 1956;6:80-6.

76. **Skinner EW, Pomes CE.** Alginate impression materials. J. Am Dent Assoc 1947; 35:245-56.

77. **Smith RA.** Secondary palatal impressions for major connector adaptation. J Prosthet Dent 1970;24:108-10.

78. **Stackhouse JA.** Relationship of syringe tipdiameter to voids in elastomer impressions J Prosthet Dent 1985;53:812-5.

79. **Stackhouse SA, Harris WT, Mansour YM, Hagen SV.** A study of bubble in a rubber elastomer manipulated under clinical conditions. J Prosthet Dent 1987;57:591-6.

80. **Steffel VL.** Relining removable partial dentures for fit and function. J Prosthet Dent 1954;4:496-509.

81. **Stevenson R.** Mixing bowl traction strips. J Prosthet Dent 1986; 56:515.

82. **Stewart KL, Rudd KD, Kuebker WA.** Clinical removable partial prosthodontics, The C.V. Mosby company, St. Louis, London, 1983, p.381

83. **Sturdevant CM, Barton RE, Sockwell CL, Strickland WD.** The art and science of operative dentistry, 2nd ed, the CV Mosby Company, St. Louis, AITBS, Delhi. 1989 p. 458

84. **Sykora O.** Readers round Table. J Prosthet Dent 1988;59:388.

85. **Tjan AHL.** Effect of contaminant on the adhesive of light bodied silicones to putty silicones in putty wash impression technique. J Prosthet Dent 1988;59:562-7.

86. **Vahidi F.** Vertical displacement of distal extension ridges by different impression techniques. J Prosthet Dent 1978;40:374-7.

87. **Vieira DF.** The forces that oppose the withdrawal of impressions. J Prosthet Dent 1950; 10:536-44.

88. **Walter Hoffman.** Axthelm History of dentistry quintessence publishing Co., Inc. 1981 Germany p. 252.

89. **Wang HY, Lu YC, Shau YY, Tsou D.** Vertical distortion in distal extension ridges and palatal area of casts made by different techniques. J Prosthet Dent 1996;75:302-8.

90. **Watt DM, MacGregor AR.** Designing partial dentures, Wright England, 1984. p. 153

91. **Wilson EG, Werrin SR.** Double arch impressions for simplified restorative dentistry. J Prosthet Dent 1983;49:198-202.

92. **Winstanley RB.** A modified impression technique for patient with interdental spacing. J Prosthet Dent 1982;47:107.

93. **Yoder JL, Thayer KE.** A rubber base impression technique for fixed partial dentures. J Prosthet Dent 1962;12:339-46.